WEB3

CW00330228

GET READY FOR THE NEXT DIGITAL REVOLUTION: METAVERSE. NFT, DAO, DEFI FOR BEGINNERS

By

William D. Guzman

processes, or directions contained within is the sole and utter responsibility of the recipient reader. Under no circumstances will any legal responsibility or blame be held against the publisher for any reparation, damages, or monetary loss due to the information herein, either directly or indirectly.

Respective authors own all copyrights not held by the publisher.

The information herein is offered for informational purposes solely and is universal as so. The presentation of the information is without a contract or any guarantee assurance.

The trademarks that are used are without any consent, and the publication of the trademark is without permission or backing by the trademark owner. All trademarks and brands within this book are for clarifying purposes only and are owned by the owners themselves, not affiliated with this document

Sommario

PART 1: GLOSSARIO

INTRODUCTION

The internet is a fast-expanding technology. We see new advances and inventions to improve and enrich our digital experience almost every day.

The World Wide Web has gone through various stages, which experts classify. Web 1.0 was the first stage, which lasted until 2003. According to analysts, the next stage, known as Web 2.0, was prominent until 2010. Web 3.0, the most recent stage, is currently the most talked-about topic.

Cryptocurrency appears to be similar to a casino from the outside. Its inner workings show something even more intriguing: **The blockchain.**

Institutions first helped human societies to merge when they permitted two strangers to do business by acting as trustworthy third parties. On the other hand, the internet displaced intermediaries in favor of dominant digital platforms.

Blockchains now use cryptography to eliminate the need for intermediaries. They use code to create self-governing economic networks that render dictators obsolete and scams impossible. The introduction of blockchain networks heralds the arrival of a "new" Internet: **web3**.

This book will assist you with obtaining web3 in a single step.

It reveals:

- ✓ What is web3, why do we need it, and how does it function?
- ✓ What role do cryptocurrencies like Bitcoin and Ethereum play in this?
- ✓ What is the best way to get started with web3?
- ✓ What are the examples of web3 use cases? DeFi, NFTs, DAOs, social tokens, play2earn, and the Metaverse are examples of web3 use cases.
- ✓ What is Metaverse?
- ✓ What does it mean by "decentralization," and what are the advantages?
- ✓ What are tokens?
- ✓ What is a wallet (it's not just a wallet but your identity online)

It's a thoughtful attempt to debunk the blockchain as a tool for bettering society.

You'll learn why web3 is the next big thing if you let me be the one crypto friend you listen to. You'll be concerned not because it will make you wealthy but because it has the potential to move civilization ahead, making us all wealthier in the process. In addition, you did not miss the boat. It's still early in the morning.

WHAT IS WEB 3.0

Web 3 focuses on moving power away from huge tech companies and toward ordinary consumers.

Web 3 – sometimes known as "Web 3" or "Web 3.0" – is a term you've probably heard recently. It simply refers to the internet's next version, which encourages decentralized protocols and strives to lessen reliance on huge digital corporations like Youtube, Netflix, and Amazon.

Web 3.0 ushers in a new age for the internet. It is a collection of technologies that will seek to alter the current state of the internet. Web 3.0 has no broadly acknowledged definition yet, but it is an advancement that will make the internet considerably smarter and more accessible to users. It's a constantly developing technology that uses breakthroughs to improve how people use the internet. It will be simpler, but it will also make the internet much more useful to people.

Data-driven and Semantic Webs powered by machine learning are the emphases of Web 3.0.

The ultimate objective of Web 3.0 is to create more intelligent, connected, and open websites.

MAJOR TECHNOLOGICAL APPROACH OF WEB 3.0

Many new technologies are expected to be incorporated into Web 3.0. Artificial intelligence is one of them. The web is predicted to become smarter and capable of a greater level of logic comprehension with the help of artificial intelligence. This implies that search engines will be able to learn about their users' habits and use that data to provide customized results to everyone. Self-learning systems will be used, and they will be able to use higher reasoning.

The semantic web is another major technological approach. This means that machines will be able to grasp natural language more easily, and we will be able to write words in our language. For example, it is projected that search engines will begin to use natural language in the future, and we will need to type in inquiries in the same way we speak in our daily lives. Search engine results will become much more focused when you search for any topic, and you will no longer see any irrelevant results.

Web 3.0 is projected to be more focused on the individual. Instead of focusing on their brand and image, websites will

14

focus on giving visitors a personalized and personalized experience. Users will be able to access related information on multiple web pages in seconds since the information from all over the web will be linked. In addition, social networking will gain traction, and it will become an important aspect of how information is presented on the internet.

Web 3.0 is also expected to take advantage of 3D technology to provide users with a more lifelike experience. We can anticipate three-dimensional websites with features such as avatars to enhance the user experience.

Overall, Web 3.0 will revolutionize how we seek, exchange, and create content on the internet. It'll take the internet to a whole new level.

WHAT IS THE METAVERSE

You've probably heard the hype: The Metaverse will alter your way of life.

The Metaverse refers to digital domains where people will gather to work, play, and hang out. It is a concept for the next phase in the internet's growth. The term has been evolving and rebranding itself, and it is likely to do so in the future. Some internet venues will be immersive 3D experiences that will necessitate the use of high-tech equipment to enjoy.

Technology hype cycles come and go. The Metaverse may fade away before it's fully realized. However, enthusiasm for the idea continues to increase. It's become a buzzword in the gaming, non-traditional retail, and retail industries.

Roblox, Minecraft, and other world-building games are all thrown into conversations about the Metaverse. Microsoft's anticipated $69 billion acquisition of ActivisionBlizzard was portrayed as part of a metaverse expansion. Facebook has rebranded as Meta, indicating the social network's desire to be a forerunner in the new world.

Virtual reality, augmented reality, and 3D computing all use ideas from older technologies. The present uptick in interest is merely the most recent peak in a years-long push to make these breakthroughs more widely available.

What's changed is a shift in mindset, meaning that the internet has to be reimagined. How far-reaching such reforms will be is anyone's estimate. After all, the Metaverse's road map is still in its infancy. It's not apparent if it'll be finished on time.

What is clear is that if there is money to be made, major firms will engage. Qualcomm, Nvidia, Valve, Epic, HTC, Apple, Microsoft, and Meta, are all working on new ways to interact online. As a result, you'll hear more about the Metaverse in the next years.

WHAT IS DECENTRALIZATION

Advances in the digital world are gradually shaping people's daily lives, businesses, academic institutions, corporations, and governmental entities. However, the hypothetical administrative systems used in this regard have not evolved simultaneously. One of the finest methods for resolving this is the decentralized implementation of blockchain technology. It's widely regarded as a platform that can answer various pressing issues, including digital identification, asset and data ownership, security, and future decentralized decision-making.

DECENTRALIZATION?

In the blockchain, decentralization means moving power and decision-making from a centralized entity (a person or organization) to a dispersed network.

Members of decentralized networks are discouraged from exercising control or command over one another in ways that might jeopardize the network's functionality.

WHAT IS THE SIGNIFICANCE OF DECENTRALIZATION?

Decentralization is not a novel concept. When creating a technology setup, three basic network architectures are centralized, decentralized, and dispersed.

While decentralized networks are widely used in blockchain technology, a blockchain application cannot be classified as decentralized or not. Perhaps decentralization should be extended to all aspects of a blockchain program on a sliding scale. More noticeable and pleasant support may be achieved by decentralizing the management of and access to assets in an application.

Decentralization usually comes with a few drawbacks, such as cheaper exchange rates. However, such setbacks' increased security and services make them worthwhile.

DECENTRALIZATION'S ADVANTAGES

1. Allows for a trustless environment to be created.

It is not necessary to trust other members in a decentralized blockchain system. This is because every network member possesses a duplicate or exact copy of the information contained in a distributed record. If a member's record is altered or polluted, most others in the network will disregard it.

2. Data recovery is improved.

Corporations share information with their partners regularly. Thanks to a decentralized information repository, each stakeholder has a timely and shared perception of the facts. As a result, this information is updated regularly and stored in each party's information vaults, where it may resurface when it is needed to be transferred downstream. When information is updated, it creates opportunities for data loss or incorrect information to reach the workplace.

3. Reduces the severity of deficiencies

In frameworks with an excessive reliance on explicit personnel, decentralization can reduce degrees of deficiency. These weak spots could lead to big problems, like not being able to do what was promised or giving useless help because of tired assets, occasional blackouts, blockages, a lack of good incentives for good service, or fraud.

4. Dispersion of assets has been optimized.

Decentralization can also assist in simplifying asset dispersion so that promised services are delivered with greater execution and consistency and a lower risk of a catastrophic letdown.

Blockchain decentralization is an information foundation that keeps a library of resources and trades through a peer-to-peer network. Unlike earlier peer-to-peer networks, a blockchain does not duplicate the transferred value. Ownership, agreements, products, and other information are all considered "resources" in addition to currency and transactional data.

However, taking everything into account, credit has been transferred from one member to another.

WHAT ARE TOKENS

In technical terms, a token is just another synonym for "cryptocurrency" or "crypto asset." However, it has taken on a few more specific meanings depending on the context. The first step is to define all cryptocurrencies that are not bitcoin or Ethereum (even though they are technically also tokens). The second refers to digital assets like DeFi tokens built on the blockchain of another cryptocurrency.

Tokens can be used for many things, from making decentralized trades easier to selling rare items in video games. They can all be bought, sold, and owned, just like any other cryptocurrency.

In cryptocurrencies, the word "token" is used a lot. Since all crypto assets can be thought of as tokens, you might hear Bitcoin called a "crypto token" or something similar. But the word has come to mean two different things, and both are used so often.

Any cryptocurrency besides Bitcoin and Ethereum is called a "token" (even though they are also technically tokens). Since Bitcoin and Ethereum are the most well-known cryptocurrencies, it's helpful to have a word to describe all the other coins.

If you're interested in decentralized finance, you'll come across this term (or DeFi). DeFi coins like Chainlink and Aave run on top of, or leverage, an existing blockchain, most often Ethereum's. The alternative, more particular use of "token" is to designate crypto assets that run on another cryptocurrency's blockchain.

Tokens in this category assist decentralized apps, from automating interest rates to selling virtual real estate. They can, however, be owned or sold in the same way as any other cryptocurrency.

WHAT IS THE SIGNIFICANCE OF TOKENS?

Given how frequently you'll see the term while researching cryptocurrencies, it's helpful to be aware of some typical meanings. However, in addition to the broad definitions in the preceding section, certain crypto-assets have the term "token" in their name. Here are a couple of such examples:

Tokens from DeFi

A new universe of cryptocurrency-based protocols has evolved to replicate traditional financial-system operations (lending and saving, insurance, and trading) in recent years. These protocols provide tokens that can be used for several purposes while also being traded and held like any other cryptocurrency.

Tokens for Governance

These special DeFi tokens give their owners a say in the future of a protocol or program that doesn't have boards of directors or any other central authority (due to its decentralized nature). For instance, everyone who uses the popular savings system Compound gets a token called COMP. People who have this token can help decide how Compound is updated. If you have more COMP tokens, you get more votes.

Tokens that aren't fungible (NFTs)

NFTs show who has the right to own a unique digital or physical asset. They can be used to make digital products harder to copy and share. They have also been used to sell one-of-a-kind digital goods, like rare gaming items, or distribute digital art in limited quantities.

Tokens of security.

Security tokens are a new form of asset that aspires to be the cryptocurrency equivalent of traditional securities such as stocks and bonds. They are most commonly used to sell stock in a corporation or other industries (such as real estate) without the assistance of a broker. This is similar to how shares or fractional shares are sold on traditional exchanges. Major corporations and startups are investigating security tokens as a possible replacement for traditional financing techniques.

CRYPTOCURRENCIES

Cryptocurrency, also called crypto or crypto-currency, is any digital or virtual currency that uses encryption to keep transactions safe. Cryptocurrencies are maintained by tracking transactions and creating new units via a decentralized method.

WHAT DOES A CRYPTOCURRENCY MEAN?

Cryptocurrency is a way to pay for things online that banks don't need to verify transactions. It's a peer-to-peer system that lets anyone send and receive money from any location. Cryptocurrency payments don't exist in the real world. Instead, they only exist as digital entries in an online database that keep track of specific transactions. Cryptocurrencies are kept in digital wallets. A public ledger keeps track of all the transactions you make with cryptocurrency funds.

Cryptocurrency refers to the fact that transactions are verified through encryption. The goal of encryption is to keep things safe and secure. This means that complex coding is needed to store and send cryptocurrency data between wallets and public ledgers.

Bitcoin was the first and is currently the most well-known cryptocurrency. It was made in 2009 and has been around since. Many people are interested in cryptocurrencies because they want to trade them to make money. Sometimes, speculators drive up the prices because they want to make money.

HOW DOES IT WORK?

Cryptocurrencies are built on the blockchain, a public, decentralized database that keeps track of all transactions and is updated by the people who own the currency.

Coins are made through "mining," which involves using a computer's processing power to solve hard math problems to make coins.

If you own cryptocurrency, you don't have anything real. You have a key that lets you send a record or unit of measurement from one person to another without the help of a reliable third party.

Blockchain technologies are still in their early stages for financial applications. The technology could trade bonds, stocks, and other financial assets.

EXAMPLES OF CRYPTOCURRENCIES

There are thousands of cryptocurrencies. Some of the best-known are:

BITCOIN:

Bitcoin was the first and is still the most widely traded cryptocurrency. It was created in 2009 and is still the most popular. Satoshi Nakamoto made the currency. Most people think that Satoshi Nakamoto is a fake name for an unknown person or group.

ETHEREUM:

Ethereum has its own money, Ether (ETH), also known as Ethereum. It was created in the year 2015. After Bitcoin, it is the second most popular cryptocurrency.

LITECOIN:

This money is like bitcoin, but it has moved faster to add new features, like faster payments and ways to make more transactions possible.

RIPPLE:

Ripple was started in 2012 as a technology for distributed ledgers. It was made by people who worked with several banks and other financial institutions. Ripple can track transactions with cryptocurrencies, but it can also track transactions with other currencies.

"Altcoins" is the name for cryptocurrencies that are not Bitcoin.

WHAT ARE NFTS

A non-fungible token. That doesn't help things much.

The term "non-fungible" simply indicates that it is one-of-a-kind and cannot be replaced. A one-of-a-kind trade card cannot be duplicated. You'd get something else if you switched it for a different card. A bitcoin is fungible, which means you can exchange one for another and get the same thing.

It's worth noting that NFTs may be used in various blockchains. (Some have previously done this.)

IN THE NFT SUPERMARKET, WHAT SHOULD YOU BUY?

NFTs have gotten a lot of attention as a digital development of fine art collecting. However, the present excitement is centered on using technology to sell digital art. Anything digital can be used as an NFT (drawings, music, even your brain being downloaded and transformed into an AI).

But, sure, this is when things start to get strange. A digital file, including the art with an NFT, can be copied.

NFTs are created to provide you with something you won't find anyplace else: control over your work (the artist can still retain the copyright and reproduction rights, just like with physical artwork). In other words, anyone may buy a Monet print in terms of tangible art collecting. On the other hand, the original may only be possessed by one individual.

Beeple's shot sold for $69 million at Christie's, $15 million more than Claude Monet's painting Nymphéas. That Monet will be appreciated as a tangible thing by whoever acquires it. A copy is almost as excellent as the original for digital art.

So, individuals have formed communities around their things for a long time, and now it's happening with NFTs. Pudgy Penguins, a group of NFTs, is one of the most well-known groups centered around the tokens, but it is far from the only one. CryptoPunks, one of the early NFT projects, has a fanbase, and other animal-themed ventures, such as the Bored Ape Yacht Club, have their subculture.

Of course, communal activities are reliant on the community. Pudgy Penguin and Bored Ape owners use Discord to communicate, trade memes, and complement one other's Pudgy Penguin Twitter avatars.

WHAT ARE DAOS

DAOs are a strong and secure way to work with like-minded individuals worldwide.

Think of them as a business born on the internet and is owned and run by all its members. They have built-in safes that can only be used with permission from the organization. Decisions are made with the help of proposals and votes so that everyone in the company has a chance to speak up.

A CEO can't just spend money on whatever they want, and a dishonest CFO can't mess with the books.

The DAO's spending limits are encoded into its code, and everything is on display.

WHAT ARE THE OBJECTIVES OF DAOS?

Starting a business that involves money and finance demands a high level of trust in the individuals you're collaborating with. However, trusting someone you've only ever interacted with on the internet is difficult. When it comes to DAOs, you don't have to trust anyone else in the group. You have to trust the DAO's code, which is fully public and can be checked by anyone.

- ✓ This opens up a lot of new ways for countries to work together and coordinate. The majority of the time, it's flat and democratized.
- ✓ The majority of the time, it's hierarchical.
- ✓ A majority must approve any changes of members.
- ✓ Changes can be sought from a single party, or voting can be offered, depending on the arrangement.
- ✓ Without a trusted intermediary, votes are counted, and the outcome is applied automatically.
- ✓ If voting is permitted, votes are counted internally, and the vote results must be handled manually.
- ✓ The offered services are handled automatically and decentralized (for example, distributing philanthropic funds).
- ✓ It requires human intervention or centrally controlled automation, both susceptible to manipulation.
- ✓ Every action is entirely transparent and open to the public.
- ✓ The majority of activity is private.

EXAMPLES OF DAO

Here are examples of how a DAO may be used to aid understanding:

- ✓ A charity can accept membership and donations from individuals worldwide, and the organization can decide how the funds are spent.
- ✓ A freelancer network – you may establish a group of freelancers who pool their funds to pay for office space and software subscriptions.
- ✓ Grants and ventures - you may create a venture fund that collects money and votes on which businesses to sponsor. In the future, money returned might be redistributed among DAO members.

HOW DO THEY WORK?

A DAO's smart contract is its backbone. The contract specifies the organization's policies while also safeguarding its funds. If someone tries to do something that goes against the rules and logic of the code, it will not work. Rules can only be changed once the contract is live on Ethereum. No one can spend the money without the organization's permission, as written into the smart contract. Rather, the group makes decisions, and payments are sent automatically after the votes are approved. This eliminates the requirement for a centralized authority in DAOs.

This is possible because smart contracts on Ethereum are tamper-proof once they go live. You can't unilaterally modify the code (the DAOs rules) without others seeing it because everything is public.

WHAT IS THE DEFI

Distributed ledgers, like those used in cryptocurrencies, are at the heart of decentralized finance (DeFi). Banks and other financial organizations no longer control financial products and services under the new system.

As technology develops, third parties are removed from financial transactions, known as decentralized finance (DeFi). DeFi's infrastructure and regulations are currently being developed. Developing apps is made easier with DeFi's combination of stablecoins, software, and hardware.

There are several advantages to DeFi for many users, including these:

- ✓ Banks and other financial organizations no longer charge fees for accessing their services, and you save your funds in a secure digital wallet instead.
- ✓ You don't need to get permission to use it if you have an internet connection.
- ✓ You may send and receive money in seconds or minutes.

GETTING TO KNOW DECENTRALIZED FINANCE (DEFI)

It's helpful to know how centralized finance differs from DeFi to comprehend decentralized finance and its operation.

CENTRALIZED FINANCE

Banks, companies whose ultimate goal is to make money through centralized finance, keep your money. Third parties which enable money transfer between parties exist in the financial system, each demanding a fee for their services. You bought a gallon of milk using your credit card. The merchant sends information about the customer's credit card to an acquiring bank for each transaction.

- ✓ The network stops the charge and asks your bank for the money.
- ✓ Your bank accepts the charge and sends it to the network.
- ✓ The acquiring bank then sends it to the merchant.

Reduced transaction times and increased access to financial services are two of DeFi's goals. Merchants must pay for the opportunity to accept credit and debit cards, so each entity in the chain is compensated. Loan applications can take days to process, and banks may not be accessible while on the road, so it's best to avoid other financial transactions.

DECENTRALIZED FINANCE

By allowing people, merchants, and corporations to perform financial transactions using developing technologies, decentralized finance eliminates intermediaries. This is accomplished using peer-to-peer financial networks that employ security protocols, connectivity, software, and hardware innovations.

If you have an internet connection, you can lend, trade, or borrow using software that records and verifies financial transactions in distributed databases. All users' data is collected and aggregated by a distributed database, which can be accessed from multiple locations and then verified using a consensus method.

As part of the decentralized finance movement, this technology is being used to make it possible for people worldwide to access financial services, regardless of their identity or where they are located.

Governments, law enforcement, law enforcement, and other organizations protect people's financial interests. People can better manage their finances with DeFi applications, which give them access to their wallets and trade services. Although outside parties can no longer control decentralized finance, anonymity cannot be guaranteed. The entities that access your transactions can track them down even if they don't bear your name.

WHAT IS THE PROCESS?

The blockchain is powered by decentralized apps (dApps).

There are blocks on the blockchain where transactions are recorded, then validated by other users. After a transaction has been verified, the previous block is closed and encrypted. The information from the previous block may be included in the current block afterward.

The information in that block "chained" the previous blocks together in each succeeding block, giving the blockchain its name. Since the information in previous blocks cannot be modified without affecting subsequent blocks, there is no way to edit a blockchain. This and other safeguards bolster a blockchain's security.

Debt financing is one type of product offered by DEFI Financial Products.

DeFi relies heavily on P2P (Peer-to-Peer) transactions. A P2P DeFi transaction occurs when two individuals agree to transfer bitcoin for products or services without an intermediary.

As an analogy, consider how centralized finance lends money. You would have to fill out an application with your bank or another lender. To use the lender's services, you'll have to pay interest and service costs.

DeFi peer-to-peer lending does not exclude the possibility of interest and fees. Because the lender may be located worldwide, you'll have a wider range of possibilities.

Using a decentralized financial application (dApp), you would enter your loan requirements, and an algorithm would match you up with other borrowers who fulfilled your criteria. One of the lender's conditions must be met before you may receive your loan.

The same blockchain technology distributes the money to the lender when you use your dApp to pay someone. As soon as the consensus confirms the transaction, the funds will be released to you. As soon as you meet the loan agreement terms, your lender can begin collecting payments from you.

An example of digital currency is DEFI CURRENCY.

Despite the fact that technology is always evolving, it is hard to foresee how or whether existing cryptocurrencies will be employed. DeFi is a bitcoin transaction processor. The concept relies on Stablecoin, a cryptocurrency linked to a fiat currency like the US dollar. BitCoins are a type of digital currency. DeFi is being built with cryptocurrency in mind. Hence, Bitcoin isn't so much a part of it as a component.

Definitively, DEFI will be around soon

It is still in its infancy when it comes to decentralized finance. For beginners, it is unregulated, which means that the ecosystem continues to be plagued by infrastructure failures, hacks, and frauds.

Each financial jurisdiction has its own set of laws and regulations, and this is how the existing legal structure works. DeFi's capacity to conduct transactions across national borders will pose a huge regulatory problem. If a financial crime crosses borders, protocols, and DeFi applications, who is in charge of investigating it? How would the rules be enforced, and who would be in charge? The decentralized finance ecosystem's open and distributed character may also impact financial regulation.

There are other issues with system stability, energy usage, carbon footprints, system upgrades, maintenance, and hardware failures.

Before DeFi may be successfully exploited, several issues must be solved and improvements made. If DeFi succeeds, banks and businesses are almost certain to find a way to obtain access to the system, if not to control how you access your money, then to profit from it.

DECENTRALIZED FINANCE: WHAT IS THE POINT?

It is the goal of DeFi to eliminate the need for third parties in financial transactions.

WHAT IS A WALLET (IT'S NOT JUST A WALLET BUT YOUR IDENTITY ONLINE)

A digital wallet also called an electronic wallet, is a mobile app that lets you send and receive money. It saves your payment details and passwords securely. You can use your device to pay for purchases after entering and storing your credit card, debit card, or bank account information. These apps allow you to pay with your phone while shopping, eliminating carrying your cards around.

On mobile devices, digital wallets allow you to store funds, perform transactions, and track payment history. A digital wallet may hold all of your financial information, and some even allow you to keep identification cards and driver's licenses.

There may be a digital wallet in a bank's mobile app or a payment app like PayPal or Alipay. People in financially underserved areas can now use digital wallets to obtain financial services that they previously couldn't.

In addition, digital wallets can contain the following information:

- ✓ Gift certificates

- ✓ Cards for members
- ✓ Cards of loyalty
- ✓ Coupons
- ✓ Purchasing Tickets for an Event
- ✓ Tickets for flights and public transportation
- ✓ Reservations at hotels
- ✓ License to drive
- ✓ Cards of identification
- ✓ keys to a car

HOW DOES IT WORK?

Digital wallets are apps that use the features of mobile devices to make it easier for people to get financial products and services. Digital wallets make it almost unnecessary to carry a physical wallet because they store all of a person's payment information safely and compactly.

Digital wallets use your phone's wireless features, like Bluetooth, wifi, and magnetic signals, to send the payment information to the point of sale that can read the information and connect via these signals. Mobile devices and digital wallets currently use the following technologies:

✓ Quick response codes (QR codes) are information-storing matrix bar codes. You use the camera on your device and the scanner in the wallet to make a payment.

✓ Near field communication (NFC) is a way to connect and share data between two smart devices. It does this by using electromagnetic signals. For two devices to connect, they must be within an inch and a half of each other (4 cm).

✓ Magnetic secure transmission (MST): This is the same method that magnetic card readers use to read your card when swiping it through a slot at a POS. This encrypted field is made by your phone, and the point of sale can read it.

Your device transmits the card information you've put in your wallet and selected to use for a transaction to the point of sale terminal connected to payment processors. The money is then channeled through the credit card networks and banks via processors, gateways, acquirers, or other third parties participating in credit and debit card transactions.

When you purchase by holding your phone over a point of sale, you're using your digital wallet to complete the transaction.

Companies like Bitpay have made cards that let you use cryptocurrencies to pay for things. The Bitpay card turns cryptocurrencies into dollars at the current market value. You can then use your wallet to pay for things with the dollars. Bitpay debit cards can be added to digital wallets like Apple Pay and Google Pay.

THE DIFFERENT TYPES OF DIGITAL WALLETS

There are a variety of digital wallets to choose from. The following are a some of the most well-known:

- ✓ ApplePay Cash App
- ✓ Google Wallet (a service provided by Google).
- ✓ PayPal Venmo Samsung Pay
- ✓ AliPay
- ✓ Walmart Pay is a service that allows you to pay
- ✓ Dwolla
- ✓ Vodafone-M-Pesa

Most wallets use various techniques to set themselves apart from their competition. For example, Google's digital wallet service lets you contribute money to your phone or device's wallet. The money can then be spent at Google-accepting businesses in-store and online.

In contrast, Apple has formed a strategic agreement with Goldman Sachs to issue Apple credit cards and develop its ApplePay services.

THE BENEFITS AND DRAWBACKS OF DIGITAL WALLETS

One of the best things about digital wallets is that they cut down on how much personal and financial information you carry.

You won't need to carry real cards or a wallet if you save everything in your digital wallet—no, there's the risk of a card slipping out of your wallet or being left in an ATM slot. Furthermore, you are unable to lose your complete wallet.

Businesses can use digital wallets and people worldwide to accept payments and get money.

You don't need a bank account at a branch for digital wallets. Instead, you can put your money in an online-only bank that offers financial services to people who don't have or don't have enough money in a bank account. This makes it easier for more people to be financially included.

If you buy a digital wallet from a company that hasn't been checked out or hasn't built up a good reputation, security could be an issue. If you lose your phone and aren't password-protected, you risk allowing someone else to access your finances. Additionally, you may wish to shop at local shops that do not yet have a point of sale that accepts this technology.

PART 2: LET'S EXPLAIN THESE TERMS

WHAT IS THE METAVERSE, AND WHAT IS THE TECHNOLOGY BEHIND IT

The Metaverse is a weak idea. People usually describe it as an online place where people can interact, work, and play as avatars. It's a step up from the internet. These places are shared and always available. Unlike a Zoom call, you can use them again and again. Many people think that the 2D digital worlds of Roblox, Minecraft, and Fortnite, where players can interact, are already the Metaverse. The first version of Second Life's Metaverse has been around for almost 20 years. (It's going to look different.)

No one agrees on whether you'll need VR or AR to get into the Metaverse, but the two will work together. Mark Zuckerberg, CEO of Facebook, Satya Nadella, CEO of Microsoft, and other supporters want to create a more immersive experience by combining different technologies, such as virtual reality headsets, mobile devices, personal computers, and servers connected to the cloud.

New VR and mixed reality headsets are likely to come out from Meta, Sony, Apple, etc. These futurists envisage creating a 3D virtual environment that may be accessed through a headset or augmented reality glasses. This means that these headsets will work with anything.

METAVERSE AND VIRTUAL REALITY

Some people think that augmented reality overlays exist in a metaverse and the real world. One idea that keeps coming up is that the Metaverse will be a virtual world that looks like our real world. There will be digital neighborhoods, parks, and clubs. They might be in one virtual world or several.

Investors are already spending a lot of money on virtual plots of land. Barbados has shown interest in opening an embassy in the Metaverse, which shows how popular the idea is.

Some people don't believe that the Metaverse will be everything Zuck, and others say it will be. Many people point out that you need big headsets to get to the most interesting parts of the Metaverse. (The person who made the Playstation called them "simply annoying," and a top executive at Meta called their headset "wretched.") They say that Big Tech hasn't figured out how to stop bullying, spreading false information, and hate speech on the internet yet. They think it will be hard to handle these issues in a world where things get even more.

WHAT WOULD LIFE IN THE METAVERSE BE LIKE?

The deluxe Metaverse, the one that needs a headset, is based on the idea of a digital universe that surrounds you from all sides. You'll have your customizable avatar and digital assets with titles that will probably be kept on a blockchain. Some believe you'll acquire digital land and construct online homes where you may entertain your friends (or at least their avatars).

Even though this may seem strange or crazy, bets have already been made on the value of digital land. A Canadian company called Tokens.com has put almost $2.5 million into Decentraland, which is a 3D world platform similar to Geocities or Second Life. (In Decentraland, you use an ethereum blockchain token to buy things.)

Some people imagine a more fluid meeting. There are now simpler metaverse experiences like Roblox and Fortnite.

These games aren't as immersive as the Metaverse Zuck talks about, but they give you a good idea of what's to come.

All the things we do on the internet already show how the Metaverse could grow: social media, gaming, Zoom telepresence, VR and AR splashes, and more. Expect to try many different ways to make it all work in fun or helpful way.

The Metaverse, according to tech CEOs like Mark Zuckerberg and Satya Nadella, is the internet's future. It might also be a video game. Maybe it's a more obnoxious, unsettling version of Zoom? It's difficult to say.

On the other hand, there is a lot of marketing hype surrounding the idea of "the metaverse" (and money). Apple's move to limit ad tracking has harmed Facebook's bottom line, particularly putting it at risk. A future where every person has access to a virtual wardrobe is hard to distinguish from Facebook's plans for profiting from selling virtual clothing. It's not only Facebook hoping to profit from the Metaverse hype.

What Does "Metaverse" Really Mean?

If you want to get a better sense of what the term "metaverse" means, try this experiment: Replace "the metaverse" with "cyberspace" in a statement. Instead of a single type of technology, the phrase refers to a wide-ranging shift in how humans interact with technology. Often, you won't see much of a difference in how the word is used. There is a good chance that the name will become dated even if the original technology is widely adopted.

"The metaverse" refers to virtual worlds that can be accessed via PCs, game consoles, and even mobile devices such as the sections of Fortnite. Virtual and augmented reality (VR and AR) are not the only ways to access these areas. Virtual reality and augmented reality are two technologies that companies refer to when discussing "the metaverse."

THE CURRENT STATE OF THE METAVERSE

It's impossible to define the Metaverse without defining away the present. Our current MMOs include virtual concert halls, video chats with people worldwide, online avatars, and a variety of commerce options. So, if you're going to market these things with a fresh perspective on the world, they need to offer something new.

Let's say you've talked about the Metaverse for a while. Someone is bound to bring up fictional works like Snow Crash, which coined the term "metaverse" in 1992, or Ready Player One, which depicts a virtual reality world where everyone can do everything from work to play to shopping.

In your mind, replace "the metaverse" with "cyberspace." In the vast majority of cases, the underlying meaning remains mostly unchanged.

Some individuals argue that this type of excitement is more crucial to the notion of the Metaverse than any one technology. People who advocate for something like NFTs, such as digital ownership certificates, are also interested in the Metaverse. Even if NFTs are horrible for the environment and most of them are based on public blockchains, which have a lot of privacy and security concerns, a tech firm can argue that they'll be the digital key to your Roblox home, then boom! Meme-buying has become an important element of the internet's future infrastructure, which may have increased the value of your bitcoin.

While it's easy to draw parallels between the early internet and the proto-metaverse and assume that things will get better and move forward more smoothly, this is not a given. VR and AR technology may never be as user-friendly as today's PCs and smartphones, so there's no assurance that people will want to hang around in a virtual workplace without legs or play poker with DreamWorks CEO Mark Zuckerberg.

In the history of technology, many disastrous investments have been made. Speculative investors have been drawn in by the term "the metaverse," coined by Facebook after it changed its name from "Facebook" to "Facebook." 3D TV, Amazon's delivery drones, and Google Glass all indicate that a paradigm change is not imminent merely because a lot of money is being invested in a certain area

However, this does not imply that nothing exciting is in store. VR headsets like the Quest are more affordable than ever before, and consumers have finally abandoned their costly desktop and console systems. It is growing easier to develop video games and other virtual environments.

Another amazing tool for digital artists is photogrammetry, which converts images or films into digital 3D models.

Tech companies can benefit from futuristic thinking, but not all do. Selling a phone is wonderful, but selling the future is preferable. A true "metaverse" would be limited to a few great VR games and digital avatars in Zoom conversations, not most of the internet.

WHY YOU SHOULD BE READY FOR

THIS REVOLUTION

Imagine an island off the coast of Britain with a population of 11 million people, rising at a rate of at least 1 million people per year, where English is the primary language, and the vast majority of the population is between the ages of 17 30. Could you afford not to market to this group? Would you be interested in marketing to them?

Surprisingly, such an island, or perhaps I should say, a group of islands, does exist, but not off the coast of the United Kingdom, but in Cyberspace. Its name is Second Life, and it is just one of a slew of virtual worlds springing up around the world. It is developed by its citizens and hosted by a hosting Lab, essentially democratic environments that adapt and change as citizens build and develop their landscapes and form their communities of interest.

This is the place to be if you want to be part of the next big thing in customer experience: the virtual experience. Why? Because the internet is rapidly becoming a platform for active customer engagement and virtual involvement.

Consider Facebook, YouTube, and Myspace, which are part of a networking Mega-trend capable of swiftly spreading positive and negative word of mouth.

But it's the coming of these 3D worlds where you can choose a character (an avatar) and walk around a virtual world at the touch of a button, conversing with other virtual characters - all genuine people signing on and participating. For the most part, it's free!

Consider how relaxing it would be to sit at home and meet a new group of friends from the comfort of your armchair—talking to someone who shares your interests but lives in China, Brazil, the United States, or Leicester! Attending an online business conference where you can interact with and see the other participants. Attend and dress up for a post-conference online party at one of the virtual discos, and, of course, purchase your avatar's clothing with virtual currency such as Linden Dollars from a virtual store.

The current wave, like many new enterprises - remember the Internet 15 years ago? - is only now beginning to be incorporated into the broader marketing mainstream. The conventional arguments that it's only about cybersex or full of geeks miss the point entirely since today's geek or student is tomorrow's customer or company leader - remember those 11 million accounts! Furthermore, 3D technology and virtual communication will almost certainly mix with the current Internet version 2.0. So, rather than thinking of this as a bizarre phenomenon separate from the commercial activities on the internet, consider it to be the next logical step.

If your bottom line isn't where you want it to be, the issue could be directly under your feet, above your head, and all around you. It's your office, and it might be the source of your company's demise.

Efficiency is the number one rule of survival in today's economic climate. The onerous expense of maintaining physical offices and all that goes with them - landline phone systems, servers, energy, furnishings, cleaning services, the list goes on and on - is one of the major efficiency drains for many firms.

The growth of technology and communication has transformed how businesses run and people work in the Digital Age. The business will become leaner, smarter, and more responsive in the future.

For many organizations, having all employees work the same hours and under the same roof is a luxury that has passed them by. As a result, telecommuting is no longer a nice benefit for employees but a critical component of a more competitive business strategy.

While getting rid of your physical workplace can save you a lot of money every month, the advantages don't end there. According to the Telework Research Network, corporations would gain over $230 billion in productivity if every American who could work from home did so even half of the time.

Furthermore, being virtual can improve the quality of your workforce by allowing you to hire the finest candidates for the position, rather than just those who reside close enough to commute to your office.

What's even better?

As virtual businesses become more common, the stigma historically associated with the lack of a centralized business operation swiftly dissipates. Today's sophisticated clients no longer value luxurious offices. Their budgets are limited, and they need to get the most bang for their buck. Customers want to know that they're investing in results rather than paying for conference rooms and copiers and talent and experience rather than footing the cost for your overhead.

Are you ready to become a part of the revolution? Is your business ready to become virtual? Before you hand over the keys to your office, you should consider several aspects to ensure that it is a suitable fit for you, your employees, and your clients.

THE VIRTUAL WORLD: THE NEXT DIGITAL REVOLUTION

Imagine a consumer pushing a button on your website that takes them on a virtual tour of your store or other business (maybe even a manufacturing plant in the planning stages) and allows them to speak with a virtual assistant about financial goods or something completely different.

You could also teleport (redirect) them to your site or area in Second Life if they already have an avatar (virtual self).

Consider attending a virtual conference to meet new people—alternatively, a discussion group. Alternatively, you may create your micro Metaverse or an innovative new site geared at a certain interest group.

Consider the 100 million or so users of social networking sites like MySpace. Members will be pushed towards 3D worlds like Second Life soon.

There's little doubt that some people will find this a difficult play to comprehend, but it's all second nature for those who grew up with video games.

Several firms have already studied the potential and begun developing virtual environments to advertise their brands. Take Amazon.com, for example, where you can search and shop in a virtual bookstore linked to and integrated with the website.

Other examples from The Official Guide to Second Life include:

- ✓ Starwood Hotels showcased its new hotel chains in Second Life.
- ✓ Toyota Motor Corporation debuted a driveable Scion xB vehicle.

- ✓ Adidas, which sells virtual versions of its a3 Micro ride shoes, has a virtual version of its a3 Micro ride shoes.
- ✓ Hipster Apparel is an American clothing retailer that has launched a virtual store to offer virtual replicas of its products.
- ✓ Harvard Law School, which provides a Second Life-related course
- ✓ Duran Duran has created a website to promote their shows and new bands.

And why not go the other way and provide an outstanding virtual store (maybe for a clothing line) before going out in the real world, whether through bricks and mortar or clicks and mortar?

This isn't only a place for tangibles like clothing and books; it's also a place for companies, as we've seen with the emergence of Second Life banks and professional services consultants. Companies like Amazon.com will be leaders in this field, but this is just as ideal an environment for a utility or financial services provider, whether for PR or guidance.

The truth is that businesses on Second Life are utilizing it for the same reason they use the internet: to spread word of mouth. Furthermore, because the primary goal is interaction, many specific interest communities are already forming, providing an ideal foundation for feedback, focusing your messaging, and getting research input.

So, what can you expect if you're interested in getting in early? How do you approach this market the right way and take advantage of the internet's power?

This is a great place to apply Customer Experience concepts. Of course, some customers will only be interested in playing Online-Poker. Still, as the Metaverse evolves and 3D technology mixes with the current Web environment, the companies with the first-mover advantage will be the victors if you were early enough to catch the wave.

HOW EMERGING TECHNOLOGIES ARE CHANGING THE GLOBAL ECONOMY'S FUTURE

The world is on the verge of a digital revolution, with technology disrupting everything from using appliances and gadgets to how we conduct business.

The digital economy is rapidly expanding around the world. The modern digital economy is characterized by the emergence of new asset classes and the digitalization of conventional assets. Blockchain, artificial intelligence (AI), the Internet of Things (IoT), and 3D printing are all new technologies that are helping to fuel this expansion.

The assets in the new technologies have the potential to dominate the global economy in the future. The blockchain, for example, has virtual coins and tokens whose popularity has exploded in a short period.

Big Names are Making an Appearance in the Game

The blockchain allows users to conduct transactions more securely and time-efficiently. Many significant technologies and financial businesses, including IBM, Oracle, JP Morgan Chase, and Boeing, have expressed interest in the blockchain's characteristics. For example, IBM has partnered with Stronghold, a financial technology firm, to establish Stronghold USD, a dollar-backed cryptocurrency. This virtual currency is an example of how customer trust in a traditional asset (in this case, the US dollar) is leveraged to back a digital asset.

There are numerous examples of corporations merging two new technologies to create future-oriented offerings. Boeing has announced a partnership with SparkCognition, an artificial intelligence company, to create blockchain-based traffic management systems for uncrewed air vehicles.

The Game-Shifter

Traditional assets, such as currencies, aren't the only ones that may be tokenized. To supply security tokens, the new market can use the intrinsic worth of a wide range of assets. The blockchain has the potential to distinguish security tokens from traditional securities. The use of smart contracts on the blockchain reduces transaction costs by eliminating the need for a middleman. The blockchain's usefulness can greatly impact the traditional banking system. Because all assets are liquid, readily available, and divisible, it may also eliminate the necessity for money as a means of exchange.

Many markets have already been impacted by automation and artificial intelligence. Machines in the manufacturing sector have taken over many occupations traditionally performed by people. Trading algorithms have surpassed human traders.

A New Framework is Required

Traditional decision-making models and procedures are no longer viable in today's fast-changing market. We must establish a new framework to stay up with emerging advancements like DAO, AI, VR, P2P, and M2M. Put another way; we need to look beyond Munger's Mental Models and concentrate on digital models like network theories and exponential growth models.

The digitization of our economy is happening at a breakneck speed. We'll have a better idea of which advances will dominate the new web 3.0 economy as time goes on, but it's apparent that this is a global economic revolution.

DIFFERENCE BETWEEN WEB1, WEB2, AND WEB3 (EVOLUTION OF TECHNOLOGY)

In 1990, the internet was nothing more than a collection of linked computers. Tim Berners-Lee built the web as its initial application.

Web1 is a "hyperlinked information system." as the name implies. Users can access a massive library of data gathered on a screen from machines all across the network by clicking around linked text and images.

Does this ring a bell?

Thirty years later, three billion users are now connected to a considerably larger, quicker, and more pervasive web powered by gigantic data centers 30 years later. The clicking has essentially remained the same.

A few decades ago, the internet was regarded as a specialist tool, utilized almost exclusively by academics. Five years later, mass adoption became a reality when browsers like Mosaic and Microsoft Internet Explorer were developed.

Surfing was at its peak throughout these years. You'd get in touch with someone. A photograph might take years to download.

The default search engine was Altavista. Nobody had considered web design before.

The first website was:

- ✓ Decentralized — Powered by ordinary computers and ordinary people.
- ✓ Open-source – Anyone can create on the internet.
- ✓ Read-only mode — Most users were readers because publishing content necessitated some technical knowledge.

Web1's original ethos was symbolized by its decentralized architecture. Without the authorization of central gatekeepers, anyone might publish any information to everyone in the world.

WEB2

In ten years, the Wild West had coalesced around victors like YouTube, Facebook, and Twitter, attracting massive numbers of users and talent in a black-hole-style fashion.

Anyone may now publish online for the first time. As obstacles fell away, so did the number of people and their usage. There was something for everyone on the internet.

Three major shifts in the backend shaped web2 as we know it today:

Smartphones allow us to go from spending a few hours per day at our computers to be "always connected" Apps and notifications rule our lives.

We've come a long way from exchanging photographs with pals to getting into strangers' vehicles. — Friendster, Myspace, and Facebook all urge people to put their faces forward and break out of their shells. These tools make it easier to create, collaborate with others, and propose new products.

Thanks to Amazon, Google, and Microsoft, it's now easier to start a business online. If you don't have the money to buy and maintain your hardware infrastructure, you may now rent it from massive data centers worldwide at a fraction of the price. The internet is becoming increasingly centralized. It's a collection of closed systems that interact with one another.

Big Tech is sucking you dry.

The central platforms exploded like mushroom clouds as we suddenly got access to more people, ideas, and technologies than our brains knew what to do with, consolidating network effects into monopoly power.

As the number of users on a network grows, it becomes exponentially more valuable. You use WhatsApp to communicate with your buddies. Mom joins WhatsApp to communicate with you. Dad joins WhatsApp to communicate with his mother. Before you know it, everyone on the planet is using WhatsApp. You are unable to depart.

Take it or leave it: WhatsApp announced in February 2021 that it would collect extra user data for business. You and millions of others have sworn to stop using the app in favor of more secure alternatives. It turns out that escaping the network's gravitational attraction isn't enough. While many people now use Signal and Telegram to communicate, only a few have fully abandoned WhatsApp. You still want to speak with Mom, and she wants to speak with Dad.

Customer value is a direct result of network size in this digital era. Users are unable to exit. Startups are unable to compete. Media, developers, and producers are forced to cooperate. The lure of the network is too great.

We pay the price for being locked in not only in dollars but also in personal data and content. To be mined, sold, and fed back into hidden algorithms that divert our attention to get us to give more. This is hidden behind the guise of "free" and "improving user experience."

The internet of today is an oligarchy. Their market capitalization is equal to your self-expression. Our interactions, searches, content, media, and data are controlled by Google, Apple, Facebook, and Amazon (GAFA). The open forum has morphed into a gated community.

Websites and applications that leverage user-generated content for end-users are Web 2.0. Many websites today use Web 2.0, which emphasizes user interaction and cooperation. Web 2.0 also aimed to improve network connectivity and communication channels.

Web 3.0 differs from Web 2.0. It focuses more on using machine learning and AI technologies to give appropriate material for each user rather than just the content that other end users have provided. Web 2.0 allows users to contribute and collaborate on on-site content. However, Web 3.0 will most likely delegate these tasks to semantic web and AI technology.

TRENDS

It took more than ten years to move from Web 1.0 to Web 2.0, and Web 3.0 is predicted to take just as long, if not longer, to deploy and transform the web completely.

However, some believe that Web 3.0's defining technologies are already being developed. Using wireless networks for smart home appliances and the Internet of Things (IoT) is a good illustration of how Web 3.0 affects technology.

When looking at the evolution of the web from Web 1.0 to Web 2.0, it can be seen that the way websites are developed and how users interact with them is predicted to alter dramatically in Web 3.0.

"user-friendly" is returning, defining, in digital terms, what is loosely referred to as Web 2.0 after many years and various technologies. Web 2.0 is the most recent version of the internet. It's the ideal market for consumers and advertisers: now it's Adsense rather than DoubleClick; live blogs rather than static webpages.

The lack of user-friendliness that prevented us from mastering our VCR is comparable to what made Web 1.0 the stuff of unmet potential for online advertisers. Before the advent of newer, more user-friendly technology that we see today on the internet, players in Web 1.0 couldn't help but fumble through the trial-and-error process.

Web 1.0 wreckage litters the inaugural Internet bubble of five years ago. It's funny how many of us classify as nearsighted animals; some even believe the internet has limitations. At the very least, I assumed the internet had some limitations. And so did Wall Street. On the other hand, stock prices do not reflect development; rather, they reflect the ultimate word on investor mood.

The term 'Web 2.0' comes from a technology conference of the same name in 2004. It loosely describes the internet's second coming, a rebirth to replace the trash strewn about after the dot.com bubble burst in 2000-2001. Web 2.0 is distinguished by its improved usability for the user. Everyday users, rather than a restricted group of online gurus, are involved in its development.

For some naysayers, the Web 2.0 phenomenon is a contentious subject. While some attempt to define a Web 3.0 (overkill?), others believe that a crucial feature of the Web 2.0 paradigm, the democratization of the Internet domain, is what will ultimately fail us.

The online encyclopedia Wikipedia.org, although being one of Web 2.0's darlings due to its open-source success, also attracts criticism for the same reasons. It's the most useful of resources, and its popularity has exploded. Furthermore, users frequently return to the site after using it, demonstrating its high quality. It's in the top 25 most popular websites, which isn't terrible for a site that doesn't advertise. Google, for example, is a free website whose success isn't reflected in apparent advertising.

The services provided by Wikipedia are completely free. Its content is likewise entirely contributed by users. On the other hand, Wikipedia is vulnerable to the flaws of amateur contribution since amateurs built it. A hoax performed by a coworker recently incorrectly listed a Tennessean as being linked to both Kennedy assassinations on Wikipedia. Such shenanigans are conceivable due to the open-ended structure.

The open-source format's attractiveness is undeniable, but there are more difficulties than sporadic performance or the occasional office joke. Wikipedia suffocates the possibility of a viable and authoritative online encyclopedia built professionally due to its open nature. The market will not exist. As things are, many people use Wikipedia.org as a starting point before moving on to more reliable sites.

I'm not sure how to soothe the pain that individuals may experience due to the impending wave of online democracy, but I can assure you that fighting the tide of Web 2.0 will be fruitless. Still, some people are set in their ways and will try to impose their view of how things should be on the internet's developing freedoms and new realities.

The "The Performance Marketing Standard" article in the Fall 2005 issue of Revenue Magazine contained an interview with a marketing executive from OgilvyOne North America, the traditional advertising powerhouse Ogilvy & Mather's internet advertising branch. The executive utilized business jargon to liberally slather promotional clichés across the interview, nearly completely ignoring the topics addressed to her. In other words, she's treating the Internet audience the same way she would a broadcast audience. That's so 1.0, isn't it?

An unintended meaning enmeshed in her jibber-jabber is impossible to overlook. In response to a query on why ad agencies have been reluctant to adapt to the changes brought on by online advertising, she claims that agencies, particularly Ogilvy, have been "leading the revolution" and must push to understand their targets. I don't blame her any more than I blame politicians whose work requires them to make statements that make me uncomfortable regularly. On the other hand, her statement exemplifies how the initial assault, or Web 1.0, failed to storm the castle of Internet success.

You make the push and attack your target on TV or radio. Advertising agencies tried to identify and push their commercials onto their targets in the early days of the internet. They looked at user behavior and devised a strategy to meet them with "effective" advertising. Non-contextual pop-ups and banners were deemed "effective." Web 1.0's aggressive poppers did not produce quality CTRs; instead, they generated enraged surfers. This strategy failed and continues to fail now. A Web 2.0 mindset recognizes that we can only accommodate our "target" to the extent that they allow it. When the user locates the advertiser, the smart advertiser is ready. It's as easy as searching.

ICMediaDirect.com began as an entirely online advertising firm. Like any other online advertising agency, we tailor our efforts to the whims of Internet users. In recognition of Web 2.0's reality, we acknowledge that the Internet user, as the public's voice, is the driving force behind "the revolution" Here, the regulations are different. Our conversations with clients do not consist of plotting to persuade the general public of anything. Our objective is to make advertisements as enticing and accessible to web searchers as possible, fully complying with the Web 2.0 flow.

Going with the flow here refers to accepting that the searcher, or internet user, is in charge of driving the boat. We have no intention of bombarding children with advertisements. Rather, we get the advertiser ready for the user. And happily, the Web 2.0 searcher is more prepared for e-commerce than the Web 1.0 searcher. The proof is in the pudding: SEM and SEO work.

Even though the content is free, it can still be free. The internet is, above all, a vast network of individuals who are increasingly successful in circumventing the conventional gatekeepers. The key is to entice users to come to you when they are ready and on their terms. That is what we excel at. Web 2.0 is here to stay, for better or worse. The smart will figure out how to get involved. VCRs are available for the remainder.

PROBLEM SOLVED BY WEB3

Cooperative governance frameworks for previously centralized products can now be widely distributed via Web 3. Tokenization can take the form of a joke, artwork, social media posts, or even conference tickets for Gary Vee.

The gaming industry is an excellent example of the paradigm shift. Large Web 2 companies like Meta and Ubisoft use Web 3 to build virtual worlds. As a result of a recent patch, their favorite weapon has been thrown out of whack, and they're constantly complaining about it. As a result of Web 3, players are able to invest in the game and vote on its future direction. By allowing players to become the sole owners of the items they acquire through non-fungible tokens (NFTs), the gaming industry will be reshaped.

THE WEB 3.0'S CHARACTERISTICS

AI, the semantic web, and omnipresent characteristics could be incorporated into Web 3.0. With the help of artificial intelligence, end-users will be able to access information faster and with greater accuracy.

Using artificial intelligence, a website should be able to find and present the information that it believes a particular user will find most relevant and useful. Social bookmarking as a search engine can produce better results than Google because the results are websites that people have voted on.

AI. With AI, results could be distinguished between legitimate and fraudulent, similar to social bookmarking and social media, but without negative feedback. Humans can control those results.

There will be a shift from a device- or app-based approach to an artificially intelligent web, where virtual assistants will be introduced.

So that a system can understand what data means, the semantic web categorizes and stores data. An alternative approach is to have a website that can read and understand human-like the words used in search queries. The semantic web will teach a computer about the data, and AI will then use the data in this method.

The term "ubiquitous computing" refers to the idea of gadgets in a user's environment being able to interact with one another through embedded processing. There's a good chance that Web 3.0 will feature this as well. The concept of the Internet of Things is a part of it.

Machine learning and artificial intelligence are only some of the technologies employed to implement these new functionalities. Blockchain and other P2P technologies will play a larger role in Web 3.0. Other technologies, such as open APIs, data formats, and open-source software, may be used while designing Web 3.0 apps.

Critics of Web 3

Critics of Web 3 technology argue that it doesn't live up to its promises. Early adopters and venture investors retain the bulk of the shares in blockchain networks, resulting in an unequal distribution of ownership. Block Inc. CEO Jack Dorsey and a slew of venture capitalists recently had a public feud that brought this topic to Twitter.

One of the most common criticisms leveled at blockchain projects is that they are "decentralization theater," that is, projects that are only decentralized on paper. For example, private blockchains, venture capital-backed investments in blockchains, and decentralized finance (DeFi) protocols are all part of the "decentralization theater."

Although the protocols community is ostensible without a leader, distinct leaders emerge. It was pointed out by Izabella Kaminska, a former editor of the FT; blog Alphaville that Vitalik Buterin, the co-founder of Ethereum, still has considerable influence over the network:

He's a strange and contradictory phenomenon in his own right, Vitalik." At the same time, Kaminska told The Crypto Syllabus, "he serves as the spiritual leader of what is effectively a de facto headless system that he has constructed and overseen."

Things aren't much better with decentralized financial protocols. Because developing blockchains look to be arcane wizardry reserved for only the most highly specialized engineers, they suffer from voter absenteeism, depend on centralized infrastructure, and have a high barrier to entry.

Web 3 has a lot of potentials, despite its faults. Over the next decade, the general public will find out if this vision is too idealistic to put into effect.

BLOCKCHAIN IN THE REAL ESTATE SECTOR

If you're still unsure about the genuine influence of blockchain on the real estate business in the coming years, take a look at some blockchain development figures for the coming years. That way, you'll be able to comprehend the technology's capabilities completely.

In addition, a growing number of diverse businesses have begun to use this technology to increase the effectiveness of their services and save money. For example, many healthcare providers and firms in other industries have recently moved their focus to blockchain development. So, let's take a look at how blockchain distribution varies by industry:

✓ $1 billion in the banking and insurance industries

- ✓ Over $700 million in the finance industry
- ✓ $653 million in manufacturing and resources
- ✓ $642 million in retail and professional services, including real estate

As a result, real estate is one of the industries that can profit from blockchain adoption for corporate growth.

WHAT ARE THE MAIN ISSUES THAT THE BLOCKCHAIN APPLICATION SOLVES IN REAL ESTATE?

As you may be aware, the commercial real estate business has seen a decline in growth over the last several years. It's because it's had to deal with several major issues that have remained roadblocks to the market's progress. At the same time, blockchain development aids in the creation of solutions to these problems. So, let's have a look at some of the most pressing real estate issues and how blockchain technology might assist in solving them:

TRANSPARENCY IS LACKING, AND OPERATIONS ARE SLOW.

One of the most serious issues in real estate is the complete lack of transparency among contractors, which leads to corruption, fraud, and money laundering while also impeding the industry's growth.

One of the key advantages of blockchain in real estate is creating shared secure databases.

The records of leasing, purchasing, and selling transactions become public knowledge, ensuring that realtors do not trip on one other's toes. Multiple listing services, which compile property-level data from brokers' and agents' proprietary databases, are an excellent example of why this new technology is so important.

Many independent parties can utilize the blockchain-enabled database simultaneously, but only those who should have legitimate access to it. As a result, only real estate management contractors such as owners, tenants, lenders, investors, operators, and a variety of other service providers can have constant, reliable access to and the ability to edit or add needed information.

USE CASES FOR BLOCKCHAIN IN REAL ESTATE

Because real estate parties now have common access to and use the same blockchain technology, they no longer need to be concerned about data integrity.

As a result, blockchain allows all stakeholders to construct a platform for safe, transparent, and speedier communication, automation, tokenization, and real-time data access, all of which are highly appreciated in real estate.

DATA MANAGEMENT OR UNSECURED TITLES

Another issue in real estate is poor record-keeping, which is necessary for quick business procedures.

This technology now enables the disintermediation of title corporations via blockchain. All entities acquire digital IDs with blockchain that can't be misplaced or appropriated. As a result, things like property titles, liens, and finance can be kept in more transparent records.

TRANSACTIONS THAT ARE BOTH SLOW AND UNSAFE

Many real estate transactions are subject to conditions, take a long time to complete, and must be transferred safely. As a result, blockchain allows for more efficient transaction processing. A purchase-and-sale transaction, for example, could be contingent on title clearances or loan approvals. Real estate companies can use blockchain to see if transactions have been completed and met terms.

Another issue with real estate is that it needs to be safer and malware-free. At the same time, blockchain addresses this by giving a better level of data encryption protection. For example, as we built for the Extobit bitcoin exchanger project, it allows encrypting all data transactions in the database to prevent any data breach.

Another example of why blockchain is worth using is a Swedish tax agency that employs it as a solution in real estate transactions, complete with an explanation of how it works.

As a result, real estate businesses may be confident in transaction security and speed while employing blockchain technology.

REAL-WORLD APPLICATIONS OF BLOCKCHAIN TECHNOLOGY IN THE REAL ESTATE INDUSTRY

Keeping in mind the major issues confronting the real estate business, we already know that blockchain development solves these issues by enabling a level of connection and security among real estate companies that has never been feasible before.

But first, let's look at some real-world instances of blockchain applications in real estate and the commercial outcomes they produce:

1. **Real Estate Tokenization:** Increasing Real Estate Liquidity and Lowering Crowdfunding Barriers

Tokens indicate a specific number of shares in certain real estate assets that may be issued, bought, and sold using bitcoin on blockchain systems. This results in a far quicker property sale.

It's worth noting that this approach lowers entry barriers for ordinary property investors. Commercial property investing from abroad becomes more manageable as well. Simply, properties can now be exchanged on exchanges like stocks.

Due to its ability to boost real estate liquidity, blockchain has the potential to change the whole commercial property market.

Take, for example, Imbrex. Imbrex is a real estate blockchain firm that uses the Ethereum blockchain. It provides free access to a wide, rotating properties market for sellers, buyers, and agents. They even get paid for supplying the data that keeps everything working.

For example, you can invest in any listed property overseas through this site without being physically present here.

Ordinary investors may soon be able to lay a claim to assets that they would not be able to purchase via current methods, owing to commercial real estate technology businesses.

2. **Smart Contracts:** Contract Digitization and Fee
   Reduction

Smart contracts, which are extremely valuable in the
financial and banking industries, are one of the most
profitable blockchain technologies. However, the real estate
market, which deals with many transactions, can profit
progressively from this technology.

With this kind of technology in place, a property
transaction that entailed a mountain of paperwork may now
be completed online between the buyer and seller. That
transaction has a better level of transparency and security
than previously conceivable.

All of the transactions are automated, needing very little
human intervention. Once set in action, anything built on
the blockchain becomes self-executing. There is less time
and effort on the part of the principals and lower costs, and
no risk of fraud.

It's also worth noting that, according to one tenant poll,
more than 55 percent of renters are unsure about the
security of their lease arrangement. In most jurisdictions,
negotiating a rental contract might take a long time and
require legal assistance. A smart contract can be created

and executed using blockchain technology. This builds trust among all parties involved and lowers the need for expensive legal counsel.

Chromaway, a blockchain realty company that has teamed with Telia, a Swedish telecommunications giant, is an example of a blockchain realty company that leverages blockchain technology. Their goal is to eliminate all barriers to digitizing real estate contracts and mortgages. Because of blockchain security features, these types of papers can be authenticated beyond a shadow of a doubt once the technology is in place.

Even if they have no prior experience in this field, blockchain technology allows anyone to construct a smart contract. They can e-sign agreements that bind both parties to the contract's terms, making what was once a cumbersome and time-consuming process much more efficient.

3. Transaction Security and Control Secured and Faster Payments with Less Fraud

Real estate blockchain applications can also lower the danger of fraud. You can be dealing with someone seeking to buy your property, or you might be trying to buy theirs.

You don't know them. Therefore, there's no reason to trust them blindly. You don't have to with blockchain technology in place. Because blockchain eliminates the potential of anything shady happening, property transfers no longer need to go through third parties.

To date, the real estate sector has required extensive documentation and the involvement of many intermediaries. As a result, property financing and payment methods have become delayed, expensive, and opaque.

When considering how blockchain may be utilized in real estate, one of the first things that come to mind is how it can simplify payments and increase the security of real estate transactions. In both rental and purchase cases, blockchain can be used to prove that parties have the funds required for the transaction.

Financial and payment systems will soon be secure and transparent, thanks to the widespread use of blockchain technology. They will be archived so that either party can refer to them.

4. Property Management Automation Reduces Administrative Costs and Time

Blockchain technology will remove human paperwork and require various software systems in the real estate industry. All of this will be replaced with blockchain technology for betterment and up-gradation. The entire property management procedure will be effective and efficient thanks to the usage of a single decentralized application with blockchain-backed smart contracts.

5. Better Decision-Making Through Transparent Data Tracking and Analysis

Blockchain is a distributed ledger that lasts as long as the network does. As a result, all information about the property or the building's history is recorded, accessible, and transparent to all future owners and investors. The blockchain can make real estate investing more equitable for all parties involved.

Furthermore, blockchain technology and Bid Data allow for more accurate tracking of consumers and owner histories across borders and banks. As a result, the danger of default is reduced. Big-data real estate players may now more correctly evaluate data and make real-time decisions based on their findings.

Property management software development and property management technology trends may interest you.

EXISTING BLOCKCHAIN IN REAL ESTATE CHALLENGES

At this time, the most major obstacles to blockchain in real estate are technical and legal issues. Blockchain real estate ventures will not be able to take off until this technology has been thoroughly researched and mastered. The market continues to face challenges due to a scarcity of personnel with experience in blockchain development.

While some blockchain technology is now in use across various apps, some users are wary of it because it is not yet legal. Another objection is legal, as legal regulations and all states have yet to adopt blockchain. Expect blockchain to become a true staple of the real estate business once it becomes more broadly recognized and understood.

It's possible that blockchain may not become an industry standard in real estate for another 10-15 years. It is, however, on the way, and if the parts start falling into place more quickly, expect a major revamp of the industry that will benefit everyone involved.

Furthermore, many things have yet to be learned about leveraging the blockchain in conjunction with AI, Machine Learning, Big Data, and IoT technologies to answer numerous challenges in today's real estate sector.

At this time, blockchain real estate transactions may appear unfamiliar. However, they offer a wide range of practical uses in security, data accessibility, and the other aspects already discussed. As a result, you can expect this technology to completely disrupt the real estate or other industry sector in the next years.

NFTS, THE NEW DIGITAL ART

Blockchain-based NFTs are unique tokens that can only be found on the blockchain. Even though the work may only have one definitive edition, as in the case of a van Gogh painting, the work may have 50 or hundreds of identically numbered copies as in the trading card format.

Who would pay tens of thousands of dollars for something that is essentially just a trading card in their right mind?

That's part of the reason why NFTs are so cumbersome. It's hard to tell which people see them as the future of fine art collection (read: a playground for the mega-rich) or as Pokémon cards (the latter of which is more accurate) (available to everyone but the mega-rich). NFTs relating to a million-dollar box of Pokemon cards were recently sold by Logan Paul.

For example, Marvel and Wayne Gretzky have launched their own NFTs, which appear to be aimed at traditional collectors rather than cryptocurrency enthusiasts. NFTs aren't "mainstream" in the same way that iPhones or Star Wars are, but they do appear to have some staying power outside of the cryptosphere.

Some of FEWOCiOUS's NFT drops have been more successful than others, but the 18-year-old claims to have made $17 million. New York Times interviewed a few NFC industry kids, and several said they used NFTs to get used to working in a group or to make some extra money.

If you have between $1,800 and $560,000, you can sell anything digital (including articles from Quartz and the New York Times) as an NFT. Deadmau's digitally animated stickers have sold well. William Shatner has sold Shatner-themed trading cards (one of which was an X-ray of his teeth).

This one has a certain allure for me. As an NFT, can I buy someone's teeth? Deadmau and Mad Dog Jones Gross were not paid $700 for this photograph, but they contributed to it anyway.

A few people have tried connecting NFTs to real-world objects as a type of verification. To ensure the authenticity of their footwear, Nike has developed a technology called CryptoKicks, which uses an NFT system. There aren't any teeth I've found so far, though. No, I'm terrified of what I'll see.

NFTs have spawned a slew of markets where users may purchase and trade them. They include OpenSea, Rarible, Grimes's selection, Nifty Gateway, etc.

More on the kittens.

NFTs became theoretically feasible once the Ethereum blockchain included support for them as part of a new standard. Most people are probably familiar with CryptoKitties, a game that enables players to trade and sell virtual kittens.

If you're not interested in creating a digital dragon's cave full of art, NFTs can be exciting since they can be utilized in games, as demonstrated by the cats. Several games already sell NFTs as in-game currency. As an NFT, a player may be able to buy an in-game rifle, helmet, or other items that are unique to the game. NFTs may be utilized to trade in virtual real estate.

If nothing else, it's better than having virtual pet rocks.

NFT pet rocks can cost tens of thousands of dollars, yet some individuals are ready to pay for them (the website says that the rocks serve no purpose other than being tradable and limited).

Are there any drawbacks?

What are your plans if you come up with a brilliant idea for a digital sticker? Does the iMessage App Store allow it to be sold? There's no way around it. NFTs allow you to sell work that would be difficult to market; otherwise, you might be interested in them.

You may also activate a feature in NFTs that pays you a portion every time an NFT is sold or transferred, ensuring that your work is incredibly popular. You'll get a piece of the action when its value rises.

One of the most apparent benefits of acquiring art is that it allows you to support your admired artists financially. With NFTs, the same holds (which are way trendier than, like, Telegram stickers). Basic rights to use an NFT, such as the ability to publish the image online or make it your profile photo, are typically included in the purchase price of the NFT you're purchasing. Owning a piece of art backed up by a blockchain entry gives you bragging rights.

As with any other speculative asset, NFTs can be bought and sold at a profit if the value of the NFT grows over time. However, I'm a little self-conscious about mentioning it.

Do I have the ability to steal a museum and take NFTs?

Cryptocurrency thefts have occurred in the past. That's up for debate. For one thing, blockchain is more difficult to steal and resell than a picture in a museum since it maintains account of every transaction. As a result, it relies on how the NFT is kept and how much work a potential victim is prepared to put in to retrieve their assets.

In 500 years, should I be concerned about digital art?

Probably. Bit rot is a genuine phenomenon that causes images to degrade, file formats to become unreadable, websites to fail, and users to lose their passwords. When it comes to physical art, museums are shockingly weak.

Make the most of my time in the Blockchain world! Buying NFTs using cryptocurrency is a possibility.

Yes. Probably. Ethereum is accepted in a wide range of marketplaces. If you want to buy an NFT and demand any money, you may potentially do it at any time.

Is it possible that trading my Logan Paul NFTs will cause Greenland to melt because of the increase in global temperatures?

Using the same blockchain technology as energy-intensive cryptocurrencies, NFTs require a lot of electricity. Most NFTs are still tied to cryptocurrencies that generate a lot of greenhouse gas emissions, even though people are working hard to fix this issue. Some artists have opted not to sell NFTs or have canceled future drops after learning of their possible impact on climate change. Since my coworker has done an extensive inquiry, you may read this essay to obtain a whole picture. Keeping an eye on it is a good idea.

How can I keep my NFTs safe in a bunker or art cave?

Like cryptocurrencies, NFTs are held in digital wallets (though it is worth noting that the wallet specifically has to be NFT-compatible). It's also possible to store the wallet on an underground computer.

Would it be possible to find a program on NFTs to watch?

Believe it or not, there are solutions available to you! Dominion X is a television series produced by Steve Aoki based on a character from a previous NFT production. On the show's website, it is said that it will be an episodic series that will be published on the blockchain (the first short film is now accessible on OpenSea).

In addition to Stoner Cats, another product that uses NFTs as a ticketing system is Chris Rock and Jane Fonda's comedy flick Stoner Cats. A Stoner Cat NFT (a.k.a. TOKEN) must watch the current episode, which is the only one available.

LEVERAGE THE TOOLS OF THE WEB3 TO CREATE SERVICES IN THE METAVERSE

Web 3 concepts like NFTs are only a part of creating the next generation of the internet.

Recent years have seen much use of "metaverse" and "Web 3" interchangeably. There shouldn't be any confusion or disagreement between these two ideas about how we want to continue to grow the internet, even though they both point to a better future for the internet.

One can only imagine the Metaverse, named after the 1992 science fiction novel Snow Crash. It is synchronous, durable, and scaleable as a 3D immersive world with unlimited concurrent users. Many people see it as such. Most of our time will be spent working, learning, playing, and entertaining in this digitally native world.

Because the Metaverse has yet to take shape, it seems foggy and abstract. It should synthesize various iterative attempts and technological advances with no clear end in sight. Even though some engineers want the idea anchored in the style of Meta's Ready Player One-like address, the Metaverse will need everyone's involvement and engagement actually to come to fruition.

On the other hand, Web 3 addresses specific issues with the Web 2 internet through a more narrowly focused paradigm. Its focus on ownership and power subverts the Web 3 model. When platforms like Facebook and YouTube create a walled-garden ecology, people's data is harvested, and their privacy is infringed. The power to control material is hampered.

When data is built on the blockchain, it is open, distributed, and collectively owned. As a result, transactions can be carried out without intermediaries, and anyone can monetize data stored on the blockchain.

Web 3 initiatives have already resulted in remarkable new consumer behaviors, such as the ability for content creators to sell their work in non-fungible tokens (NFT), games that allow players to earn money while playing, and a community-organized investing collective (ConstitutionDAO) that raised enough money to bid on the

United States Constitution at a Sotheby's auction. These are just a few examples.

Despite Web 3's ability to alter how we store data, govern ourselves, and trade money, the latency with which blockchain transactions are cleared limits the settings and use cases it may be deployed. The idea of a truly decentralized internet is enticing, but it is also impracticable. One can argue that Web 3 is an integral aspect of the Metaverse, but that's not quite accurate.

Instead of alienating other types of contributors, realizing that Web 3 and decentralization are simply building blocks for the Metaverse opens up their opportunity.

There was protest when Meta (previously Facebook) announced their primarily AR/VR-centric metaverse vision, fearing that Big Tech would once again dominate the metaverse and force platforms to operate as a closed environment.

People overlooked the fact that Meta's innovation and attention were mostly on hardware and a 3D user consumption and input interface that, to be honest, does not exist today. Facebook is attempting to address the immersion issue, which is significant. Consider that for a moment. Many of us have been on Zoom for the past two years and are exhausted. How will we feel if we wear a virtual reality headset all day?

If we want to spend more time in the virtual world and enjoy it, we'll need more immersive, natural, and expressive virtual interfaces. Meta's advances in AR/VR and motion-sensing technology do not jeopardize Web 3 and decentralization efforts. In reality, the best-case scenario is that people begin developing Web 3 applications for emerging 3D form factors such as AR/VR and holographic projections.

Another popular belief is that Web 3 will render Web 2 obsolete. It's difficult to envisage such a scenario. Despite some of Web 2's flaws, there are still a lot of products that work better without using the blockchain. People can chat and broadcast at scale and in real-time using platforms like Discord or Twitch. Uber and DoorDash, for example, effectively queue demand and match it with supply.

Whether you like it or not, centralization is effective. OpenSea, the world's largest NFT marketplace, is essentially a centralized marketplace that just facilitates blockchain transactions. Coinbase is another controlled exchange that facilitates cryptocurrency trades. Like any other Web 2 marketplace, these intermediaries take service fees on transactions in both circumstances.

While these hybrid products do not exactly correspond with the decentralization idea, they are important "bridging products" that aid in the widespread adoption of Web 3 aspects by appealing to the general public. Meta's adoption of Stories helped it become a mainstream product for all demographics, comparable to how Snap Stories was a popular young product but struggled with adoption with older users.

It's common to think of new technologies and paradigms as revolutions. However, we can see throughout history that they tend to build on top of previous eras' foundations. Email is still a big part of our daily lives, despite being created during the Web 1 period of the internet.

There's still a lot of construction to be done. The blockchain, play-to-earn, various occupations, virtual economy, UGC [user-generated content] platforms, and scaling content creation are stepping stones. It won't be some glitzy product launch from a corporation that says, "Hey!" We've been working on this for ten years, and now the Metaverse has arrived. It'll result from several different enterprises operating in various spaces.

While the recent development of Web 3 and efforts to mainstream blockchain use cases is a big step forward in our progress toward a better internet, it is only one component, and other complementary activities should not be overlooked.

IS IT POSSIBLE TO EARN SIGNIFICANT AMOUNTS THANKS TO THE METAVERSE?

Many corporations that have jumped on the metaverse bandwagon also see a new digital economy where users may produce, purchase, and sell items. It's interoperable in the more idealized views of the Metaverse, allowing you to transfer virtual objects like clothes or vehicles from one platform to another. However, this is more difficult than it appears.

While some proponents argue that new technologies like NFTs would enable movable digital assets, this is not the case. Moving goods from one video game or virtual world to another is a hugely complex operation that no single firm can handle.

It's difficult to decipher all of this because, when you hear descriptions like those above, you might think, "Wait, doesn't that already exist?" The environment of Warcraft, for example, is a persistent virtual world where people may purchase and trade things. Fortnite includes virtual activities like concerts and an exhibit where Rick Sanchez can learn about MLK Jr. You may throw on an Oculus headset and be in your virtual home.

Is that really what "the metaverse" means? Just some new forms of video games?

In a nutshell, yes and no. To call Fortnite "the metaverse" refers to Google as "the internet." Even if you spend a lot of time in Fortnite socializing, shopping, studying, and playing games, it doesn't guarantee you're getting the full picture of what people and companies mean when the term "the metaverse." Just as Google does not develop the entire internet—from physical data centers to security layers—it does not build the entire internet.

Microsoft and Meta are only two of the many businesses working on virtual world interface technologies. Larger virtual worlds that closely match our daily lives are being built by many companies, including Nvidia, Unity, Roblox, and even Snap.

There are practical and fascinating breakthroughs in the domain of creating digital worlds. Epic, for example, has purchased several companies that assist in the creation and distribution of digital assets to bolster its formidable Unreal Engine platform. While Unreal is primarily used in video games, it is also used in the film industry and might make it easier for anyone to create virtual reality experiences.

A Ready Player One-style single unified realm known as "the metaverse" remains entirely implausible. That's partly because such a world requires companies to collaborate in ways that aren't always profitable or desirable.

A direct link to World of Warcraft from Fortnite isn't something that the game's developers are eager to implement—and that's because the raw computer power necessary for such an idea may be further distant than we believe.

Many corporations and activists use the term "metaverse" to describe any single game or platform. New terms have been coined as a result of this humiliating fact. "Metaverse" might be anything from a VR concert app to a video game. Metaverses can be described as a "multiverse of metaverses" by some individuals. A "hybrid-verse" may exist, or perhaps we live in a "hybrid world." Alternatively, these words might imply anything.

In conjunction with a Fortnite-themed mini-game, Coca-Cola has introduced a new "taste born in the metaverse." There are no regulations.

We have a hazy idea of what the Metaverse may consist of if we reinterpret terms to suit our purposes. We also know which corporations support the concept, but even after months of discussion, no one seems to have a consensus on what it should be called. If Meta has her way, it'll contain dummy residences where you'll be able to have a party for all of your pals. Microsoft appears to believe that virtual conference rooms might be used to teach new employees or to communicate with faraway colleagues.

Finally, we see her in an Avengers-style hologram during the show. She can see and hear the music, read the text that floats over the stage, and even establish eye contact with her buddy. If you're looking for a cool new product, this isn't the place to find it. Herein lies "the metaverse's" primary flaw.

In the Metaverse, How Do You Make Money?

The metaverse revolution has arrived, so the virtual gold rush has begun. This new digital frontier is garnering interest, from huge tech companies to everyday people who want to be the first to take advantage of the metaverse's limitless possibilities. There are many methods to generate and exchange value for real-world advantages due to your experience in the metaverse, ranging from taking on virtual employment to producing new forms of art and entertainment. It's also likely to be a lot easier than you think. So, here's a list of 11 methods to profit in the metaverse.

1. NFTS FORMATION

At present, the NFT mania is nonstop. Beeple's $69 million NFT auction, the Bored Ape Yacht Club, and CryptoPunk's price increase in 2021 were just some of the high-profile examples of their rise. It's possible to sell any of your innovations in open marketplaces in the metaverse, though, by converting them into non-fungible tokens (NFTs).

The metaverse will rely on avatars, mansions, fashion, collecting cards, and other digital things designed and generated by people to build and maintain open economies. Several tutorials are available online on producing an NFT, which is digital evidence of ownership and validity of a specific property, virtual or otherwise. However, for those who aren't interested in this, there are other options to consider, such as opening an NFT art gallery and selling other people's work in exchange for a cut of the profits or becoming an art broker and advising metaverse clients on how to best navigate this newly-minted NFT world.

2. REAL ESTATE

Virtual real estate has experienced a major boom recently, with plots in metaverses like Decentraland, Axie Infinity, and The Sandbox selling for millions of dollars due to the demand for digital land. There are a variety of possibilities available:

Buying virtual land or digital assets and reselling them for a greater price while pocketing the difference is known as real estate flipping.

The need for virtual real estate businesses has only grown as the number of real estate businesses has increased. Bring buyers and sellers together or give advice, and you may make a lot of money as a real estate broker in real estate.

RENTING:

You may buy a plot of land, construct a house or other property, and then rent it out, just like real life. You might also advertise on your property, particularly in a high-traffic region.

As a manager of other people's virtual assets, such as virtual music halls and land, you may monetize your metaverse real estate abilities, such as renting out your virtual property.

Building metaverse structures and plots of land may be a rewarding endeavor. Professional 3D designers will fast become one of the most sought-after professions in virtual worlds, whether for a home, a mall, or a stadium.

3. ADVERTISING

Virtual malls allow brands to open shop and advertise their products and services, including leasing and selling virtual reality ads, just like they can in the real world. Many businesses flock to the metaverse to establish a virtual presence and exploit it as a significant advertising platform. As more individuals join the metaverse, it's just a matter of time until it becomes a big advertising and marketing platform.

4. FASHION

The metaverse's first use was fashion. High-profile fashion houses like Louis Vuitton and Gucci are already experimenting with virtual apparel through their NFT lines.

Gaming has also proven to be a lucrative avenue for brand partnerships with Burberry x Blankos Block Party and Valentino x Animal Crossing initiatives.

However, starting a digital apparel line and making money in the metaverse is possible for almost anyone. You don't have to be a fashion designer to have fun customizing your avatars and helping your friends do the same.

5. EDUCATION

The potential of virtual schooling has become clear due to the epidemic compelling schools to close and millions of students worldwide to attend online lessons. The metaverse will continue to evolve into a more immersive place, allowing for more tailored and engaging education. Tutoring and educational programs will undoubtedly flourish as a result.

6. ENTREPRENEURSHIP

Entrepreneurs will thrive in the metaverse. Users can build stores and start enterprises in the virtual world rather than in the actual world. An unlimited number of enterprises may help you generate money in digital realms, whether in fashion, sports collectibles, real estate, or entertainment.

7. GAMING

In metaverse-type settings like Roblox and Fortnite, gaming is now the most popular activity. Users may gather and sell in-game assets in exchange for tokens that may have real-world worth, whether playing blockchain-based games or investing in metaverse activities (check out our earlier intro to P2E). You may also generate money in the metaverse by creating games for people to enjoy.

8. TOURISTIC TRAVEL

Virtual reality tours have provided much-needed relaxation for a large number of people. It is hoped that the metaverse will allow for the virtual replication of historical landmarks and events in the future. Tour guides and travel agents will be among the jobs produced in the metaverse tourism sector.

9. PROMOTION & HOSTING OF EVENTS

Entertainment will be crucial in metaverses such as Sensorium Galaxy, where superstar DJs like David Guetta and Armin van Buuren will perform. Thousands (if not millions) of people are expected to attend parties, concerts, and sporting events, so promoters and hosts will be just as important as they are in the real world to ensure that virtual worlds have unforgettable events.

10. PRODUCT TESTS FOR METAVERSE

As more products enter the metaverse, users who can test and provide feedback will be highly demanded. The possibilities for experimenting with digital assets are limitless with such a diverse range of industries and businesses set to enter the virtual universe.

11. VIRTUAL LABOR

Virtual workers will be in high demand to create and maintain a metaverse. 3D artists, virtual reality architects, community managers, developers, coders, graphic and fashion designers, recruiters, and metaverse content creators are just a few of the positions that will be in high demand. You'll be able to offer your services to companies like Meta or Microsoft or as a freelancer, no matter what sphere you're in.

HOW TO PURCHASE CRYPTOCURRENCIES

You might want to know how to buy cryptocurrency safely. Most of the time, there are three steps. Here's what they are:

Step one is to choose a platform.

The first step is to decide on a platform. You can use a regular broker or a cryptocurrency exchange most of the time:

Brokers who do business the old way: These are online brokers that let you buy and sell cryptocurrencies and other financial assets like stocks, bonds, and exchange-traded funds (ETFs). People know that these platforms have lower trading fees but fewer crypto features.

Places to buy and sell cryptocurrencies: There are many cryptocurrency exchanges to choose from, and each has its own set of cryptocurrencies, wallet storage options, interest-bearing account options, and other features. Many exchanges charge asset-based fees.

When comparing different platforms, think about the cryptocurrencies they offer, the fees they charge, their security features, how they can be stored and withdrawn, and any educational materials they might have.

Step 2: Put cash in your account

Once you've chosen a platform, you'll need to put money into your account before you can start trading. Even though this depends on the platform, most crypto exchanges let users buy crypto with fiat (government-issued) currencies like the US Dollar, the British Pound, or the Euro using their debit or credit cards.

People think it's risky to buy cryptocurrency with a credit card, so some exchanges don't let you do it. Some credit card companies also won't let you use your card for crypto transactions. This is because cryptocurrencies are very volatile, and it's not a good idea to risk going into debt or paying high credit card transaction fees for certain assets.

Each platform has different ways to pay and different times to deposit and withdraw money. In the same way, the time it takes for a deposit to clear depends on the type of payment. Some sites also accept wire transfers and ACH transfers.

Fees are an important thing to think about. These fees could include fees for making deposits and withdrawals and fees for trading. Fees will be different depending on how you pay and what platform you use, so do your research ahead of time.

Step 3: Buy something.

You can use your broker's web or mobile platform or exchange to place an order. If you want to buy cryptocurrencies, go to "buy," choose the order type, enter the number of coins you want to buy, and confirm the order. The same steps are used for orders to "sell."

There are also other ways to put money into cryptocurrency. PayPal, Cash App, and Venmo are all payment platforms that let customers buy, sell, or store cryptocurrencies. There are also the following options for investments:

Bitcoin trusts: A traditional brokerage account can be used to buy a Bitcoin trust. With these vehicles, regular stock market investors can buy and sell cryptocurrency.

Bitcoin mutual funds: Both Bitcoin ETFs and Bitcoin mutual funds are available.

Blockchain stocks or ETFs are another way to invest indirectly in crypto. These are companies that focus on the technology behind crypto and crypto transactions. You could also invest in stocks or exchange-traded funds related to the blockchain (ETFs).

Your best choice will depend on your investment goals and how willing you are to take risks.

HOW SHOULD YOU STORE CRYPTOCURRENCY?

After buying bitcoin, you'll need to keep it safe so it doesn't get stolen or hacked. Cryptocurrencies are usually kept in "crypto wallets," which are pieces of hardware or online software that safely store your private keys. Some exchanges have services that let you store your money right on the platform. But not every exchange or broker will give you a wallet for free.

There are many different wallet providers to choose from. Two different wallets are called "hot wallets" and "cold wallets."

"Hot wallets" are ways to store cryptocurrency that use internet software to protect the private keys to your assets.

Storage for cold wallets: Unlike hot wallets, cold wallets (also called "hardware wallets") store your private keys on devices not connected to the internet.

Cold wallets charge fees most of the time, but hot wallets don't.

WHAT CAN YOU BUY WITH BITCOINS?

Bitcoin was made to be a currency that can be used every day. Users can buy anything from a cup of coffee to a computer to big-ticket items like real estate with Bitcoin. That hasn't happened yet, and even though more and more institutions are using cryptocurrencies, major transactions that use them are still rare. Even so, crypto can buy many different things on e-commerce sites. The following are some examples:

Websites about technology and online shopping:

Like newegg.com, AT&T, and Microsoft, several tech companies accept cryptocurrency on their sites. One of the first places to accept Bitcoin was an online store called Overstock. Shopify, Rakuten, and Home Depot also accept it.

Some high-end stores accept cryptocurrency as a way to pay for expensive items. For example, Bitdials, an online luxury store, will let you buy Rolex, Patek Philippe, and other high-end watches with Bitcoin.

Cars: Some car dealerships, from mass-market brands to high-end luxury brands, now accept cryptocurrencies as payment.

Insurance:

AXA, a Swiss insurance company, said in April 2021 that it now accepts Bitcoin for all of its insurance lines except life insurance (due to regulatory issues). Premier Shield Insurance, which sells insurance plans for homes and cars in the United States, also takes Bitcoin for premium payments.

You can use a bitcoin debit card like BitPay to spend cryptocurrency at stores that don't accept it directly in the US.

Cons and scams involving cryptocurrency

Cryptocurrency crime is on the rise, which is a shame. Some scams that involve cryptocurrency are:

Fake websites are scam sites that use fake testimonials and crypto jargon to promise that you will make a lot of money if you keep investing.

Virtual Ponzi schemes: Cryptocurrency thieves make up fake opportunities to invest in digital currencies and give the impression of high returns by paying back old investors with money from new investors. One scam organization, BitClub Network, raised more than $700 million before its members were charged in December 2019.

"Celebrity" endorsements: Scammers pretend to be millionaires or well-known people on the internet and tell you that your virtual currency investment will make you more money. Instead, they steal the money you put in. They could even use messaging apps or chat forums to spread rumors that a well-known businessperson backs a certain cryptocurrency. After getting other people to buy and driving up the price, the thieves sell their investment, and the currency's value goes down.

Scams involving virtual currencies: The FBI has warned about a new trend in online dating scams in which con artists convince people they meet on dating apps or social media to invest or trade in virtual currencies. In the first seven months of 2021, the FBI's Internet Crime Complaint Center got more than 1,800 reports of romance scams involving cryptocurrency. These scams caused $133 million in losses.

If not, thieves could pretend to be legal sellers of virtual currency or set up fake exchanges to trick customers. Misleading sales pitches for cryptocurrency-based individual retirement plans are part of another crypto scam. Then there's just plain hacking of cryptocurrency, where hackers get into people's digital wallets and steal their virtual currency.

IS IT SAFE TO PUT MONEY INTO CRYPTOCURRENCY?

Most cryptocurrencies are made with blockchain technology. The way that transactions are recorded and time-stamped in "blocks" is what blockchain is all about. It's a long and complicated process, but in the end, hackers can't change the secure digital ledger of cryptocurrency transactions.

A two-step authentication process is also needed for transactions. You might be asked to enter a username and password before starting a transaction. Then you might have to give a code that was sent to your cell phone via text message.

Even though security measures exist, this doesn't mean that hackers can't get into cryptocurrencies. Several well-known hacks have caused a lot of trouble for bitcoin startups. Both Coincheck and BitGrail were hacked for $534 million, making them two of the most expensive cryptocurrency hacks of 2018.

Unlike government-backed money, the value of virtual currencies is only based on supply and demand. This can cause big changes in the market, making investors make or lose a lot of money. Also, the government has much less control over cryptocurrency investments than it does over traditional financial instruments like stocks, bonds, and mutual funds.

WHEN INVESTING IN CRYPTOCURRENCIES, THERE ARE FOUR THINGS TO WATCH OUT FOR.

Consumer Reports say that there is a risk with every investment, but some experts think that bitcoin is one of the riskier ways to invest. If you're considering investing in cryptocurrencies, these tips can help you make smart choices.

Learn about bitcoin exchanges before you invest in a partnership. Do your research, read reviews, and talk to investors who have more experience before you choose. It is thought that there are about 500 exchanges to choose from.

LEARN HOW TO PROTECT YOUR DIGITAL MONEY:

If you buy cryptocurrency, you have to store it. You can save it on an exchange or in a digital wallet. There are many different wallets, and each has its own set of pros, cons, and security features. Before you invest, you should look into your storage options, just like you would with an exchange.

Spread out your investments:

Diversification is an important part of any good investment plan, but it's especially important for cryptocurrency. Bitcoin is a well-known name doesn't mean you should put all of your money in it. There are many options, and it's best to spread out your investments by buying other currencies.

Prepare for turbulence:

Know that the cryptocurrency market is very unstable, so you can expect ups and downs. Prices will change in a big way. If your investments or mental health can't handle it, cryptocurrency might not be a good fit for you.

Cryptocurrency is very popular right now, but keep in mind that it is still very new and is seen as a high-risk investment. When you buy something new, you should be ready for problems. If you decide to join, do your research first and start putting in a small amount of money.

When most people think of cryptocurrency, they think of a strange form of money. Even though few people seem to know what it is, everyone talks about it as if they do. This article aims to bust all the myths about cryptocurrency to know what it is and what it's all about by the time you're done reading.

You might decide that bitcoin isn't for you, but at least you'll be able to talk about it with confidence and know more than most people.

Due to their investments in cryptocurrencies, many people have already become millionaires. There is a lot of money in this relatively new market.

The word "cryptocurrency" is shortened to "cryptocurrency," an acronym. How something becomes valuable in the first place is not so easy to explain.

Cryptocurrency is a digital, virtual, decentralized currency made with cryptography, which Merriam-Webster defines as "the computerized encoding and decoding of information." Cryptography makes it possible to use debit cards, bank online, and do business online.

Cryptocurrency is electricity that has been encoded as a series of complex algorithms. Banks or governments do not back cryptocurrency. Instead, it is backed by a complex system of algorithms. The fact that they are hard to hack gives them a monetary value. It is just too hard to copy the process of making cryptocurrency.

Cryptocurrency is very different from what is called "fiat money." Fiat money is money whose value is set by law or order from the government. The dollar, the yen, and the Euro are all examples. Fiat money is any currency that is officially accepted as payment.

Another difference between cryptocurrency and fiat money is a limited supply of a cryptocurrency, like silver and gold. Only 21,000,000 of these very complicated algorithms were ever made. There is neither more nor less than what there is. You can't change it by making more of it, like how the government can't just print more money without anything to back it up. A bank could also change its digital ledger, which is what the Federal Reserve will tell institutions to do to deal with inflation.

Cryptocurrency is a way to buy, sell, and invest money without worrying about government monitoring or banking institutions tracking your funds. This system could become a steady force in a world economy that is not stable.

Cryptocurrency is also a good way to stay anonymous. Criminals can take advantage of cryptocurrency just like they can take advantage of regular money. It can, however, stop the government from keeping track of everything you do and getting into your private life.

There are many different ways to get cryptocurrency. They all result from careful alpha-numerical calculations made with a complex coding tool. Bitcoin was the first cryptocurrency, and it is still the standard by which all others are judged. Other cryptocurrencies include Litecoin, Namecoin, Peercoin, Dogecoin, and Worldcoin. Altcoins is a general term for all of these. The price of each depends on how much cryptocurrency is on the market and how much people want it.

The process of making cryptocurrency is very interesting to look at. Unlike gold, which must be mined from the ground, cryptocurrency is just a record in a virtual ledger kept on computers worldwide. To "mine" these entries, math procedures must be used. Computational analysis can be used by one user or, more often, by a group of users to find specific data blocks. The "miners" look for information that exactly matches the encryption algorithm. Then it's used in the series, and they've found a problem. The data block has been decrypted when an algorithm matches up with a similar set of data on the block. As a reward, the miner gets a certain amount of cryptocurrency. As time goes on, the reward amount decreases because there are fewer and fewer coins. Also, the algorithms used to find new blocks are getting harder to understand. Finding a matched series gets harder to do with computers. Both of these things slow down the process of making new cryptocurrencies. This looks how hard and rare it is to get something valuable like gold.

Now, anyone can work as a miner. The tool for mining Bitcoin was made open source by the people who made Bitcoin. This means that anyone can use it for free. On the other hand, the computers they use are 24 hours a day, seven days a week. The algorithms are very hard to understand, and the CPU is doing all it can. Many people have computers that were built just for mining bitcoins. Miners can be the person who uses the computer and the computer itself.

Miners, who are people, also keep track of transactions and check to ensure that a coin hasn't been copied. This stops hackers from taking over the system and making it go crazy. They get paid for their work by getting more cryptocurrency every week to keep their business running.

Their cryptocurrency is kept in specific files on their PCs or other personal devices. These files are called "wallets."

Let's go over some of the words we've already learned:

- ✓ Cryptocurrency is an electronic currency. It is also known as "digital currency."
- ✓ Fiat money is any legal cash backed by the government and used in the banking system.
- ✓ Bitcoin is the first cryptocurrency and the most valuable one.
- ✓ Altcoins are other digital currencies that work similarly to Bitcoin, but their coding is different.
- ✓ Miners are people or groups who use their resources to mine digital currency (computers, electricity, and space).
- ✓ There is also a special computer that runs many algorithms to find new coins.
- ✓ Wallet: Your digital money is kept in a small file on your computer called a wallet.

Here's how you should think about the cryptocurrency system in a nutshell:

- ✓ People use their resources to mine the coins, making electronic money.

- ✓ A stable and limited way to make money: For example, only 21 million Bitcoins have ever been made in the whole history of the world.
- ✓ There is no need for the government or a bank to be involved.
- ✓ The coins are worth what people are willing to pay.
- ✓ There are many kinds of cryptocurrency, but Bitcoin is the most well-known one.
- ✓ It can make a lot of money, but like any other investment, it also has risks.

HOW TO COLLECT YOUR FIRST MONEY TODAY, PLAYING

Only the game makers made money when people played games online, while players only spent money on in-game items. But that may change because more and more gamers want rewards for how much time they spend in virtual worlds. The rise of the "play-to-earn" model, which the metaverse made possible, made this possible.

How play-to-earn games in Metaverse work is explained.

Video games have changed a lot in the last few years, just like cell phones. They have gone from being small enough to fit in your palm to being played on a computer and a wireless console. Gamers also moved from simple games to more complex ones like Grand Theft Auto, Call of Duty, and Assassin's Creed, where they spent hundreds of dollars on skins and weapons to show off to their friends and rivals.

So far, though, only the people who made these games made money, while the people who played them just spent money to have fun. But a new generation of gamers wanted something different. They wanted rewards based on how much time was spent in virtual game worlds, and the metaverse made this possible. It made the play-to-earn model popular.

How are metaverse "play-to-earn" models different from those in mobile games?

So far, most video games have had a centralized economic model, which means that developers and publishers had the rights to all in-game economy items and could give them out however they wanted. For example, if you owned skins or other items in the game, the developers could let you keep them. But if they shut down the game or took away any features that affected the item you owned, you wouldn't be able to do anything and lose the item or skin.

Also, Mark Zuckerberg says that Metaverse will come after the Internet and will change everything.

With the "play-to-earn" model, all the digital assets you own in the game are legally yours, and you can do whatever you want with them. You can even take them to other markets and sell them there.

How do ownership and the play-to-earn model work?

Play-to-earn games use blockchain technology because you can earn items while playing them by using crypto tokens, non-fungible tokens (NFTs), and staking. Many of these games will give you one of the digital items above in exchange for your time.

For example, in the game world of Axie Infinity, you can earn tokens called SLPs that you can sell on exchanges for real money or stablecoins.

You can also sell and trade digital assets like land and weapons in the form of NFTs with other players in a special marketplace. Since these items would be turned into tokens, they would be unique assets that can't be copied, and the tokens for these items would be safely stored in a distributed ledger.

You can sell and buy virtual plots in a game called Decentraland. Recently, $1.3 million was paid for a piece of virtual land on Decentraland.

You don't have to worry about the safety of these assets because they are stored on a blockchain. And you would be the sole owner of these things. Neither the publisher nor the developer would have any ownership rights.

It's important to note that no game is truly "decentralized" because the publisher still has to define, issue, and limit the asset traded as an NFT.

A step-by-step guide to games that pay you to play them:

This guide will tell you what you need to do to start playing games that pay you to play them. Different games have different rules, so we'll discuss the two most popular ones: Axie Infinity and Decentraland.

Axie for all time

To store cryptocurrencies, you will first need an ethereum-based wallet. Metamask is a good wallet that is easy to use.

Then you'll need to set up a ronin wallet. It's the place on the ronin blockchain where you can store in-game items and assets. The chrome web store is where you can get it.

After setting up your ronin wallet, go to the Axie marketplace and use your ronin wallet to create an account.

After setting up your account, go to "account settings" and connect to your Ethereum-based wallet.

To play the game, you have to buy three Axis, which are monsters in the game, from the Axie market. Due to its rising popularity, the price of Axis has gone up. It now costs about 0.15 ether (ETH), or about $300. You can also join a program that gives people money to help them go to school. These are programs started by other Axie owners who lend out their Axis in exchange for a share of the SLP earnings.

Once your Axie account is set up and you have enough Axis, all you have to do is download the game and start playing.

Decentraland

Open your web browser and go to the website's main page. When you open the page, you will see options to play with your wallet or as a guest. It is best to play with a digital wallet because it lets you enjoy the whole game.

If you choose "play with a digital wallet," you have to link the game to the wallet you want to use.

When you connect your wallet, you'll be asked to sign in. You will be taken to the page where you can change your avatar.

On the page for customizing your avatar, you will be able to change its head, body, top, bottom, shoes, and other accessories.

Choose a name for your character after you've made changes to them.

You are now ready to look around the world of decentraland.

You can walk your character to other people's land and properties to look around, or you can buy any piece of land in the game.

The play-to-earn game model in the metaverse is a new way for people to make money from their time playing video games. The model is still in its early stages, so it's hard to say how profitable it will be for players in the future.

CONCLUSION

Web3—or the next internet—is gaining traction. This new version of the web is developed using blockchain technology, which is based on the decentralisation principle and allows for more transparency. Blockchain makes it easier to use digital assets like cryptocurrency and non-fungible tokens (NFTs).

We are at a point when everyone is overestimating what can be accomplished in a year and underestimating what can be accomplished in ten. Don't only believe the hype created by macroeconomic cycles. Instead, educate yourself on blockchain technology's long-term sustainable use applications.

As an investor in Web3 startups, I'm looking for the most creative ones that are focused on usefulness and impact and follow the decentralization ideals.

Web3 Isn't Ready Yet!

Web3 is currently mostly theoretical, with a somewhat high learning curve. Anyone interested in participating must first educate themselves on blockchain and cryptocurrency technology. Not everyone wants to go through that trouble just to use a different version of what they already have, especially when apps like private browsers can help them avoid privacy worries.

There are also concerns about anonymity and censorship. Nothing would be anonymous if the whole internet operated on Web3 blockchain architecture and everything was irrevocably put onto the blockchain. That's OK for some people, but not for others who need to stay anonymous for their own protection.

In principle, no one could be stopped from using the internet, but the dissemination of damaging disinformation and hate speech would need to be managed in some manner. Because the internet is currently so lousy at regulating these concerns, it's difficult to determine whether Web3 would be better or worse.

The most important step, of course, is transferring power away from tech behemoths like Amazon and Google. According to PCMag's Sascha Segan, this is "a political problem, not a technological one." The best of Web3 will never be realized unless the government reins in or dissolves these firms.Before Web3 can be taken seriously, it must first avoid the onslaught of crypto and NFT fraud.

Crypto Lessons

In recent years, the cryptocurrency ecosystem has gone through a similar era of openness before becoming more centralized and user-friendly.

Web3-based apps rely on cryptocurrency. According to experts, it will be used in the future to tip authors, pay for virtual products, and purchase new game features.

In the early days of the business, cryptocoins were difficult to purchase. Users have to write code in order to make their own crypto wallets. Some performed well, but others transferred money to the incorrect address, misplaced keys in their wallets, and fell prey to fraud.

As time passed, developers stepped in and built user-friendly programs to purchase, store, and trade crypto, such as Coinbase and Binance.

This might be the sweet spot for Big Tech: According to Eloho Omame, managing partner of First Check Africa, a venture capital firm located in Lagos, Nigeria, a version of the web that is less centralized than what we see now but more user-friendly than what Web3 developers can give on their own.

"I think we will end up in a world in which aspects of decentralized ownership meld together to create experiences with more control than we have under existing platforms," Omame said.

If Big Tech does get a foothold in the market, Morris thinks there is still a chance for the new web to accomplish its decentralized ideal. He says big tech may be useful for a while while Web3 firms get their foot in the door and their initiatives up and running.

"In the future, what we need to do is kick these companies out," he added.

NOTES

I hope you enjoyed this book and learned something from reading it.

My aim is to open people's minds, so as to unmask the scams that chase us every day.

Please leave a review so other readers can discover this book.

Thank you

William

Printed in Great Britain
by Amazon

85139798R00092